I
BOUGHT
A
MOUNTAIN

I
BOUGHT
A
MOUNTAIN

THOMAS FIRBANK

First published in Great Britain in 1940
by George G. Harrap & Co. Ltd.

This new paperback edition published in 2022 by Short Books
an imprint of Octopus Publishing Group Ltd
Carmelite House, 50 Victoria Embankment
London, EC4Y 0DZ
www.octopusbooks.co.uk
www.shortbooks.co.uk

An Hachette UK Company
www.hachette.co.uk

10 9 8 7 6 5

Copyright © Thomas Firbank literary estate

Thomas Firbank has asserted his right under the Copyright,
Designs and Patents Act 1988 to be identified as the author
of this work. All rights reserved. No part of this publication
may be reproduced, stored in a retrieval system or transmitted
in any form, or by any means (electronic, mechanical, or
otherwise) without the prior written permission of both
the copyright owners and the publisher.

Foreword copyright © Patrick Barkham
Afterword copyright © Dafydd Morris-Jones

A CIP catalogue record for this book is available
from the British Library.

Print ISBN: 978-1-78072-560-4

Printed and bound in Great Britain by Clays Ltd, Elcograf S.p.A.

This FSC® label means that materials used for the
product have been responsibly sourced

MIX
Paper | Supporting
responsible forestry
FSC® C104740

Also by Thomas Firbank

Fiction
Bride of the Mountain (1940)

Non-fiction
I Bought a Star (1951)
A Country of Memorable Honour (1953)
Log Hut (1954)

Contents

Foreword *by Patrick Barkham*   11

1   The Purchase   19
2   The Valuation   31
3   The Prelude   41
4   The Gathering   51
5   The Lambing   65
6   The Year the Snow Came   77
7   The Fox Shoot   90
8   The Cutting   100
9   The Pig Idea   109
10   The Dyffryn Pigs   123
11   The Poultry Tragedy   135
12   The Washing   149
13   The Shearing   158
14   The Wool Sale   171
15   The Hay Harvest   182
16   The Snack Bar   193
17   The Caravan   204
18   The Dipping   218
19   The Record Walk   227
20   The Rock-Climbers   243
21   The Annual Sale   256
22   The Rams   271
23   The Hydro-Electric Set   283
24   The Renting of Cwmffynnon   297

Afterword *by Dafydd Morris-Jones*   313

Contents

Foreword

by Patrick Barkham

IN THE MIDST OF A MAJOR CRISIS, as doubts proliferate about city life, capitalism and the technological revolution, an idealistic person seeks to escape turmoil and make a living from the land. We see this story played out today in Britain as the coronavirus pandemic causes flight from city to countryside. An urban passion for farming is writ large in contemporary culture, from Amazon's smash-hit TV show about Jeremy Clarkson's foray into farming to Cumbrian farmer James Rebanks' critically-acclaimed bestsellers *The Shepherd's Life* and *English Pastoral*. Yesterday, by chance, I stumbled across the most-read article on a Welsh news website: a derelict farm with 16 acres on sale for £250,000, a tantalising price for Londoners confined to a one-bedroom flat worth twice that.

But turning a fantasy about living more 'naturally' away from the strains of urban existence into reality is not a new story. It is a recurring theme since the industrial revolution. Where Romantics such as Samuel Taylor Coleridge and William Wordsworth led, many have followed, seeking individual fulfilment in a supposedly simpler existence in a wilder place. Thomas Firbank was not a Romantic but in this book he told one of the most compelling and successful twentieth century versions of the urge to escape to the country.

I Bought A Mountain begins at a roaring pace, in the middle of a howling gale, and never lets up for its 250 pages. It is a book written by a young man in a hurry and we are swept along by the power of his dreams and his determination to realise them.

We first catch sight of Dyffryn when Firbank visits this hill farm in the heart of Snowdonia in a November gale. An abandoned lorry is blown on its side. The bonnet of his car is wrenched off by the wind. The rain is horizontal. Perhaps most unpromisingly of all, it is 1931, and the Great Depression is starting to bite. Nevertheless, a mysterious force seems to press this rough, mountainous Welsh hill farm on to the 21-year-old Firbank, who has just fled two years labour in a factory in Canada. 'I jumped out of the car, and the wind frogmarched me at a run to the back door. The door opened unasked, and I stumbled inside,' he writes.

There, he met a Welshman who wished to retire from farming Dyffryn, 2,400 acres of grazing land on the southerly slopes of the Glyders. The terrain was inhospitable, the rainfall was seven times that of London but the price for the farmhouse, two cottages, barns and land was an alluring £4,625 (£325,000 in today's money). So Firbank 'bought the mountain' although his daughter, Johanna, later said that the book's title was always ironic. No person, believed Firbank, could actually buy let alone 'own' a wild Welsh hillside.

The farm's purchase is an irresistible set-up for a story of liberation that appeals to every reader who has ever been stuck in a meagre house or stultifying office job: the story of a man braver than us (and probably possessing more money) who jumps into a drastic change-of-life with no planning, skills or experience. The learning curve is as steep as that hillside. Firbank is not simply moving to the mountains in 1932 but taking a job enmeshed in a Welsh-speaking country, culture and society. 'I was a foreigner in a land as alien to me as Tibet,' he writes. 'The language was new to me, and, more important, so was the mentality of the people.'

Firbank is not quite the ingénue he portrays. Although he was born in Quebec, his father was English, his mother was Welsh and, as he later acknowledges, he spent many holidays with his mother's friends on farms in North Wales. But he is an outsider and swiftly learns not to pretend otherwise. 'A scientist, a doctor, an artist, a soldier, a lawyer, can all be impersonated for a short while with some hope of success,' he writes. 'A farmer never. More is needed than a glossary of jargon and a studied physical expression. No one can impersonate an earthquake or an acorn, and a farmer is just as much a natural manifestation.'

A farmer is authentic because they have to be, it is not a role that can be faked. This is deeply appealing to all of us, whether reading of, dreaming about, or actually trying to farm. The atavistic allure of farming also includes the freedom that can come from being a generalist in a world where everyone's job seems to be shrinking into an ever-narrower specialism. A farmer, Firbank writes, 'is a judge of many kinds of stock, he is a veterinary surgeon, a botanist, a chemist, an engineer, an architect, a surveyor, a foreman, a meteorologist, a buyer, a vendor, and an advertising manager. And the only job in which he really fails is the last.'

Firbank does not fail in advertising farming to us. A contemporary version of this tale would probably contain more personal feelings but Firbank deftly describes the brutal business of farming and the successes and failures from each challenge flung his way. His writing is vivid and vital and we see the world as freshly as he does, and learn the fascinating complexity of farming sheep when bequeathed with infertile soils and inclement conditions.

There are thrilling tales of lambing, flooding, shearing and digging sheep from the snow. There are schemes to generate electricity and the electric atmosphere of auction day. Firbank superbly evokes the excitement of selling off his sheep every autumn. 'The long climax of bid, counter-bid, and bang of the gavel must be sustained from the first to last. What connexion is there between a farm sale, mob hysteria, religious ecstasy, yoga, the immunity of fakirs, hypnosis? There is a connexion,' he insists. For all Firbank's usual lack of sentimentality, there are also moments of great intimacy, written with delicacy and even poetry. During lambing, a first-time mother 'stares in superstitious amazement at the slimy morsel of life which has so mysteriously appeared,' before warily approaching to sniff the suspicious object. When her maternal instinct is triggered, and she licks her newborn clean, the lamb sits 'a-sprawl, his legs at impossible angles, as if pinned to him by a blind man'.

We learn alongside Firbank how his Welsh mountain flock is 'hefted' to the hill and so even though the mountain is unfenced, the sheep will not stray; we learn how when a lamb dies, its skin must be cuts off and wrapped around a twin or orphaned lamb so the bereaved mother will adopt and feed another. Most of all, though, we learn that Firbank cannot do this alone. He quickly falls in love with a local woman, Esmé Cummins, who is memorably described as having 'the face of an elf' and being 'as dainty as a Dresden shepherdess'. Fortunately this porcelain figurine in person is also strong, practical and as bubbling with ideas as Firbank. Ultimately, Firbank swiftly realises that he cannot cope without his local community – not only the expertise of his two shepherds, John Davies and his son, Thomas, but a host of neighbouring farmers.

In this way, *I Bought A Mountain* is a portrait of a lost era when farming was a communal endeavour. Firbank requires his neighbours' help every time he gathers in sheep. Forty men assist with sheep shearing, and for no payment except the food that Esmé spends days preparing in the kitchen. Firbank must lend his shepherds freely to his neighbours as well.

For all that has changed in hill farming since the 1930s, Firbank's experiences echo more modern accounts by farmer-writers such as James Rebanks. Firbank encounters salesmen flogging every kind of farm technology and tonic. He also pursues 'diversification' 80 years before it became an agricultural buzzword. He tries farming pigs and chickens as well as sheep, which ends badly, but finds success with a snack bar he builds for the tourists flocking to the Welsh mountains. He and Esmé start a trend for reconditioned shepherd's huts decades before today's boom, buying a beautiful wooden wagon which they rent to holidaymakers.

Firbank's arguments about the importance of farming are highly pertinent today. He is writing on the eve of the Second World War, when Britain produced just 42% of its own food and knew it must urgently produce more. Ever since the lifting of the Corn Laws in 1850, which allowed the importation of cheap grain from overseas, the British government has prioritised cheap food for its urban majority over self-sufficiency for the nation. Firbank's appeal for better support for farmers is the same as many current arguments, when the global crises of climate and extinction are making an irresistibly important case for Britain to produce more of its own food without wrecking this land or any other.

As with almost any book published more than 80 years ago, the ecological message – or absence of it – within these pages does not chime with contemporary thinking. Firbank pursues what almost every twentieth century farmer regarded as his duty: 'improvement' and 'modernisation'. He mercilessly killed foxes, drained 'bogs and swamps', fertilised natural grassland and 'harrowed' square miles of 'matted pasture'. Over the last century, this 'improvement' has driven wildlife from farmland, and virtually eradicated flower-rich meadows in favour of grass or arable monocultures. Occasional instances of archaic language or attitudes within these pages may also grate with 21st-century readers. Firbank counteracts the casual prejudice against Welsh 'duplicity' that was commonplace among the English of the day and his writing is suffused with respect and generosity towards Welsh people. Sometimes, however, his humour is directed against them; and he seems untroubled by the fact that his labourers will at times toil for no payment while the boss goes on foreign holidays and drives a Bentley. Firbank and Esmé break the record for the fastest ascent of the 14 hills above 3,000ft in Wales but there are plenty of times when Esmé's role as farmer's wife is simply to make lavish teas for everyone.

So Firbank's farming story is not a flawless emancipation or perfect ecological awakening but the seeds are certainly there, as what happened next reveals. Published in 1940, his uplifting tale of personal development, resilience and strong communities was well-suited to wartime. *I Bought A Mountain* was a bestseller, and has rarely been out of print since, inspiring many subsequent generations to move to the countryside.

Unexpectedly for any reader, its publication marked the end of Firbank's farming story. Before war broke out, he left Dyffryn and enlisted in the army, where the resilience honed on the mountainside served him well. He was awarded a Military Cross for bravery in Italy in 1943, later writing a book about his experiences, before moving to the Far East in 1954 to work as an engineer. During the war, he split from Esmé but left the farm in her hands. She remarried and managed it successfully for many decades. Her deep love of this mountain led her to become an influential conservationist. In 1958, after a successful campaign to stop a youth hostel being built on the slopes of the Glyder mountains, she and her husband Peter Kirby founded the Snowdonia National Park Society.

For both Esmé and Thomas Firbank, who returned to live in North Wales in the 1980s, Dyffryn became their *cynefin*, a poetic Welsh concept which prosaic English struggles to translate as a place of belonging to which a person feels an intense spiritual connection. For all the pragmatism of the farming life on show, this spiritual connection with the land is revealed on virtually every page. The profound fulfilment and freedom both author and his wife found in hard labour in a place where they enjoyed meaningful relationships with all its inhabitants remains deeply inspiring to us today.

1

The Purchase

I FIRST SAW DYFFRYN IN a November gale. As I rounded a spur of a hill to turn into the long valley the full power of the storm caught the car. An abandoned lorry, blown on to its side, half blocked the road, and as I crept past it an eddying gust swooped down, plucked at the car's hood, and ripped it backward till it streamed raggedly behind. The rain was being driven horizontally, and struck on the windscreen. It poured in torrents over the bonnet, but left me dry.

The entrance to Dyffryn valley is guarded by two lakes. The left wall of the valley is the long hump of Moel Siabod, and the right wall, higher and more rough, is the Glyders. Across the head of the valley stand Snowdon and her satellites, like maidens hand in hand, barring the way out. But on that first day wild flurries of rain and mist shut out the skylines, and the steep, rocky slopes reared upward till they were swallowed by the clouds. Every now and again the clouds were rent like parted curtains, to reveal yet higher hills, from whose every hollow and gully streamed creaming water. The wind raced like a live thing about the upper slopes; sometimes it carried bodily away a whole waterfall, so that for a moment not a drop would spill over the brink.

The surface of the twin lakes was whipped into vicious white horses. Along the shores huge boulders lay scattered haphazard, as if untidy giant children had fled for shelter, leaving their marbles where they lay.

I liked that weather. I had come from two years' imprisonment in a Canadian factory, from an atmosphere of dust and artificial humidity. During summer we automatons used to peer through the shut windows at the shafts of sunlight as they fought their way down through the smoke and chimney-stacks to spill on the dirty paving, like storm-troops on enemy concrete. And in winter the snow lay in the streets, her virginity prostituted under careless feet, so that she taunted us, like the slut she was, with the purity of her sister in the country.

The man who can strike a mean is the man who makes a dull success of what he undertakes. But the ordinary people, who move with the ungoverned swings of the pendulum, live foolishly and fully. My pendulum was at the extremity of its arc when I came to Dyffryn. The rain was a balm, the wind a caress, the wild Welsh mountains were an elemental purge. I think I had decided to buy even before the hood was blown away.

Dyffryn had been described to me by an acquaintance who had heard that the owner was on the point of retiring. The place was a sheep-farm of 2,400 acres. It lay in a long rectangle along the south slopes of the Glyders, so that its upper boundary was the height of land on my right, some 3,300 feet up, and the lower boundary was the flooding river which ran on my left, parallel with the road. There were said to be a good house, two cottages, and plenty of farm-buildings, and the price was around £5,000. I drove on up the valley, until presently I came to a stony track which led upward on my right at a slope of one in three. The track vanished into a clump of ragged trees a hundred feet above. They were the only trees in the valley, and among them showed a chimney. The stony road was drained by slate slabs buried on edge diagonally across the track. The first of these knocked off the silencer of the car. I dared not stop, because the hill was so steep and slippery, and I was now sideways to the rain, which flooded in. With spinning wheels the car ran crabwise into a yard beside the house. The gable end of the house, exposed to the westerly gales, was window-less and slated from roof to ground. I jumped out of the car, and the wind frogmarched me at a run to the back door. The door opened unasked, and I stumbled inside.

It was very dark inside. I was in a big kitchen, lit by one small window and by the flames of a kitchen range. At a deal table under the window several men sat on a long oak settle. They were finishing a lunch of cold bacon and hot mashed potatoes. There was tea on the table, with bread and squares of red Canadian cheese. The men were dark Celtic types, sharp-featured and with the modelled mouth which stamps the Gael and the Celt. I felt that my arrival had struck them dumb. Against the opposite wall stood an oak dresser laden with blue willow-pattern plates and pewter, and beside it was a grandfather clock, the hands set three-quarters of an hour fast. The kitchen range was framed in huge blue slate slabs, and along the mantel were thirty-seven Toby jugs.

A little, spruce man with a face like a withered apple sat at one side of the fireplace. He wore a homespun wool suit and very shiny leather leggings. He was the man I had come to see. Busying herself over the fire was his wife, a pleasant-faced old lady, slight of build, her face seamed by the work and worry of a lifetime. And to bar my retreat followed the sturdy kitchen help, who had concealed herself behind the back door as she opened it to let me in. I stared at the maid several times before I realised what it was that struck me as incongruous. She was wearing pince-nez.

As I shook hands with the old couple, the men at the table rose with a clatter, unanimous as Javanese dancers in obedience to a call, and swept out with their steaming mackintoshes draped round them. The maid cleared away the remains of the meal, and at once began to lay out the best china and a dish of tinned peaches.

My conversation with the old man was difficult, because, although he had a perfect command of English, he chose to speak Welsh, using his wife as an interpreter. And she had a slow, reluctant air, as of one who acted unwillingly in some Faustian deal. The old man and I sat down to eat – farm women never eat, unless secretly – and I began to tackle him on the business which was now so urgent for me. The old house was quivering under the thrusts of the wind, and the wild, remote setting had already captured my fancy, and will hold it till I die.

I had never before bought anything not labelled with a net price, and the moves and counter-moves of bargaining were unknown to me. I must have shocked the old farmer with my bludgeoning, but he defended skilfully, and while his wife translated from one of us to the other, he had time to get his breath. I could pin him to nothing definite. He would give no price, nor would he confirm that he wished to sell. At one time, I believe, he even denied ownership. The only thing he could not deny was the letter which I had sent to warn him of my visit. The envelope, soiled by much handling, was propped against a tea-caddy on the mantelshelf.

At last I became so bewildered that I finished my food and stood up to go, half convinced that a mistake had been made. But by chance I had hit upon the right move. The farmer and his wife held a quick consultation in Welsh, and the maid was sent to the back door, where she screamed into the wind. As if by magic one of the men off the settle reappeared. He was a tallish, thin individual. Then and at each subsequent time that I saw him, a drop hung from the tip of his nose.

The old man suggested that since I had come, I might as well look round. The thin man was given some instructions, and I was hustled out after him into the yard. The rain had stopped, but the mist still hung low on the hills. Beneath the mist the air was very clear, so that one seemed to look up the red-brown slopes to a thick white muslin curtain whose frayed hem was napping in the wind. I walked round to the front of the house, which was poised on a terrace of the plunging hillside, and looked down on to the valley floor. The twin lakes still boiled in foam and spray, and the river which threaded its way through the bottom of the valley was flooding boisterously over the few flat stretches.

But the shepherd was impatient. He called to a wall-eyed dog with a grey-blue coat, and turned up the mountain. I plodded behind him; I waded through soaking russet bracken, jumped swift, swollen streams, and trod carefully where turf was raised up quaking by some force of water which bubbled through buried rocks. Presently we reached the edge of the mist. The outlines of my guide became indeterminate, and the dog a flitting ghost.

About fifteen hundred feet up we came upon a high drystone wall. This wall climbed vertically from the lakes, then turned at right angles to follow the contour of the Glyders, until it swung downhill again to join the road far up the valley. It formed a vast enclosure of the lower land. It was six and seven feet high and two feet thick. In parts it ran across turf where no stones were available for building, and in other places it had been built with jigsaw artistry over masses of glacial debris. It would today cost more to build such a wall than to buy and stock the farm. On either hand it vanished into the mist, as if bent hurriedly on a secret errand.

We climbed the wall and were on the mountain proper. The angle increased, till I was using my hands as well as my feet, and the silent shepherd led on straight upward. And after another half-hour we walked abruptly out of the mist. The edge was clean-cut, and one looked back as at a lace curtain. The mist hung in a layer between a thousand and two thousand feet, and springing from it like a score of Aphrodites from the surf were many peaks.

The ground swelled up to the height of land, and beyond the

edge it was abruptly cut away, so that we stood over a drop whose floor was hidden by mist. Across the void Tryfan rose out of the vapour. Tryfan is a black conical giantess of solid rock from whose cap spring like feathers two natural stone pillars. The pillars are named Adam and Eve. They are poised over a precipice, and in a gale it is a feat to leap from the one to the other.

We had struck the boundary at the eastern and lowest edge, but westward it reared up, till two miles away it formed the peak of Glyder Fach, the limit of the farm. Beyond this again sprawled Snowdon and her retinue, their lower slopes mantled modestly in the white drapery of the mist. Their proportions were so perfect that no stranger could have guessed their height.

We turned and made towards the Glyder, the precipice was on our right, and on our left were huge grassy hollows which fell in swoops towards the hidden mountain wall. An occasional sheep snorted and stamped at the dog before bounding away, till it was swallowed up by a fold in the ground. The peak of Glyder Fach came nearer, but her towering head began suddenly to fade. Over Snowdon rolled white clouds, like breakers tumbling a corpse, and their spray reached out in tentacles across the gap towards us. And in a moment the Glyder was gone, veiled from our view, leaving us enticed as if by a wanton woman. The hills are wholly feminine in their casual caprice. Their jewels must be won by pretended indifference and enjoyed with simulated carelessness. The shepherd gave up, and turned downhill. Our world was a few yards in diameter, and in its unreality it seemed as if we stood still and the rough ground slid silently up to us and past us.

An hour and a half later I was again seated by the kitchen fire, drying my wet legs. The atmosphere was different. I wondered if I had become tuned to the vibrations of the place. The old farmer spoke to me in English, and himself pressed me to the most buttery of the muffins, to the best jam, to the currant bread. He volunteered that he would sell at once with immediate possession for £5,000, and that he would leave enough on mortgage to enable me to take over his flock of sheep at a valuation. As I only possessed exactly £5,000 this help was very necessary to me.

The sheep valuation is a custom peculiar to certain types of hill farm. Hill farms are of two sorts. There is the farm with the fenced mountain, and there is the farm with the open, unfenced mountain – such as Dyffryn. On first thoughts the fenced mountain seems more desirable. Here the flock cannot stray, nor can a neighbour's sheep crowd in to graze. But the upkeep of a mountain fence is heavy. Iron posts and heavy wires are speedily eaten away by rust. The rainfall is tremendous in the Welsh hills, and the water is driven by the wind under each flake of spelter and into every crevice in the post. Constant tarring only delays collapse, and it is a heartbreaking task to carry tar each day up to three thousand feet.

The snow is more spectacular than the rain. Packed hard against the fence by the wind, it becomes as heavy as ice, and flattens long stretches under its weight. And the sheep, too, cannot use their intuitive weather sense. In a gale they are barred by the fence from moving over on to a lee slope, and must remain cold and miserable, plastered up against the wires or buried in drifts of snow.

The only real advantage of the fenced mountain is that the ingoing farmer is independent of a valuation. He can buy sheep anywhere in the open market and turn them on to his mountain, knowing that they cannot stray. If there were no fence, the flock would be scattered in a few days all over the county. But even though the new sheep would not be able to stray, they would be ignorant of the weather signs and of the paths and sheltering places, so that a hard winter might well kill off a third of them. The open mountain carries a permanent flock which knows the boundaries. Each year the ewe lambs are kept to enter the flock, and the four-year-old ewes are sold off to make room for them. Thus there are always four generations of sheep on the mountain. Every farm has its own earmark, a combination of slits and notches on the two ears, which is not duplicated within a twenty-mile radius. And to aid more distant recognition the farm, too, owns a pitch-mark. This is a big brand either of letters or of symbols which is stamped each year on to the new-shorn sheep.

The permanent flock stays at home by a mass heredity which is rather like the intuitive cleverness of birds. Each ewe has her own beat, and will always be found near the same grassy hollow or sheltered gully. She brings up her lamb on her little range, and when in time she is sold, the lamb carries on the tradition, and eventually bequeaths the domain to her own offspring. This legacy, multiplied by a thousand individuals, ties the flock to its home.

I used to wonder what kept the original sheep of the flock on their own ground. But in those distant days labour was very cheap. Lads with dogs would walk the hills all day to shepherd the sheep within the boundary, and at night they would pen the flock in sections in the innumerable stone pens whose ruins still dot the hillsides. And gradually the need for shepherding would become less and less as finally the flock ceased to stray. The heredity which springs from this early shepherding is still strong. Sheep from adjacent farms naturally mingle along the common boundary, but if a dog is sent along the dividing-line the sheep make for their respective sides like hairs parting before a comb. The outgoing farmer does not give away this laboriously acquired flock instinct. He knows that one dare not stock the unfenced mountain from the open market, and he levies blackmail in the shape of an acclimatization value. Hard times have reduced this premium, until today it is about 15 per cent added to the market value of the sheep.

I asked to see the buildings, and the old man led me first to the old farmhouse, which stands immediately behind the present one. I stooped to enter under a shaped headstone, and found myself at once carried back four centuries. I was in an old Welsh manor. At one end of the long, low room an immense fireplace filled the wall. Acres of peat must have been cut to keep the grate fed, and the tremendous chestnut beam which held up the chimney was blackened with the smoke of generations. The ceiling was upheld by beams of chestnut a foot thick, mortised one into the other and pegged. Upstairs were three rooms, divided by rough-hewn carved wood screens, and the rafters were pegged to the purlins with a skill long dead. The heavy roof was of clumsy slates quarried off surface rock.

The quarry is near by. I have cleaned the lichen from the face of the rock, and strange inscriptions are exposed. There is a Bardic inscription in cuneiform letters, many initials with dates long past, and addresses whose ruins now stand as monuments. And in one place is the crude outline of a ship of the Stuart period; a shadowy artist has recorded a visit to Portmadoc, then a thriving shipyard, sixteen miles away. The chestnut logs tell of a vanished forest. Often as we cut peat or dig drains we come across chunks of twisted wood, the bones of dead trees.

From the old house the farmer and I walked down the hill to the road, and I was shown the two cottages, which are sturdily built of stone and roofed with the purple Caernarvon slate, which is the best slate in the world. The cottages were let to quarrymen, for the farm-hands slept in Dyffryn house, but I determined to have my house to myself and to let my men live in the cottages. The three sets of farm-buildings were made in the same solid way, and the cattle lying on their beds of bracken were warm and contented as they chewed the short, sweet upland hay.

I drove home in a dream that night, a prince on the threshold of his kingdom.

A few days later I met the old farmer at a solicitor's office. Within a few minutes we had settled at £4,625, and enough money was left on mortgage to enable me to meet the bill for stock after the valuation. Under the benevolent guidance of the lawyer I wrote my cheque, each letter a weighty seal. My eggs were in one basket. True, it was a very big basket, but much of it was as yet unseen. The old man took me out and gave me some whisky. He wished to take me at once to see the lowland farmers who were wintering yearlings for him, and as soon as we started off in his car I was glad I had had the whisky.

On a hill farm it is necessary to send away the ewe lambs for their first winter. Unlike the lowland men, we do not put a ram to our lambs, because it is imperative to give the stock a good start if they are to survive the rigour of mountain winters. Thus the ewe lambs spend their first winter down in the mild lowlands, unhampered by the burden of gestation. It is said that ewe lambs do best on new pasture, and I soon saw how well the wintering farms had been chosen. Some farms had as few as twenty-five yearlings, one had as many as eighty, but on none of them had sheep been kept during summer, and cattle had grazed down the grass to a height handy for sheep to tackle. Many fields were newly seeded down, and some of the farms had fields of rapes, into which the lambs would be turned during the dead weeks after the New Year. The lambs are sent away from early October to early April, and had only been down a few weeks when I was taken round. I noticed that none of them were turned yet into the best pasture, and learned that they had to be broken to it slowly, used as they were to sparse mountain grazing.

As I was introduced to the various farmers, I found out that they had wintered lambs from the same flock for year after year. The lowland man is supposed to look after the stock as if it were his own, and he is not paid for any that die. The conscientious farmer thus retains the confidence of the hillman, and is assured of sheep each year. I saw over three hundred yearlings that day, and agreed to pay the arranged price when the time should come to fetch them home. The price averaged ten shillings a head. Before I parted from the old farmer that night we arranged a date for the valuation.

I was worried as I drove away after saying *au revoir* to the old man. I had certainly plunged into the water without testing it first with my toe. I knew nothing of farming in general, and would find this particular type hardest of all to understand. For its methods were traditional secrets, and were not set forth in books for the inquiring students. And I was a foreigner in a land as alien to me as Tibet. The language was new to me, and, more important, so was the mentality of the people. I wondered whether I should ever be able to probe their thoughts, and to persuade them to accept me. For bit by bit I was realizing the importance of neighbourly relations in such a place as Dyffryn. A dozen or more men were needed to gather the sheep off the mountain, and farms near by sent their men to help, knowing that Dyffryn would in turn help them. I wondered what would happen if I were boycotted; the place would be unworkable, and I supposed that I should have to clear out at a heavy financial loss. And as an example of my vast ignorance, I even considered the possibility that my sheep might be stolen wholesale. I did not realise the difficulty of stealing mountain sheep. Representatives from several farms help at gatherings, and theft is thus impossible in the pens, while it would be almost impossible for one or two thieves to drive a bunch of mountain sheep down from the hills. And I did not realise, either, that a sheep's ear-mark is as conspicuous to a Welshman as if the owner's name were in luminous letters along the ewe's back. But the most reassuring point of all did not then occur to me. Sheep-stealing would be an unthinkable act for a genuine Welsh farmer. A bishop would be as likely to rob a gas-meter.

I was also worried about the casual way in which the sheep I was to buy were to be turned back to the mountain. It seemed to me that they might go off anywhere. There was nothing to stop them. It was later that I learned how easily Welsh sheep break out of any enclosure, but they are tied to their own mountain by a heredity which is many times as strong as the best fence. And I was puzzled too over more immediate problems. None of the men at Dyffryn were staying on, except one lad. And the more I thought of the labour problem, the more I felt that I should need some sort of working bailiff. Even in those early days I deplored that unfortunate type the Gentleman Farmer. Kipling should have written of him. But I could not expect to plan work for my men when I knew absolutely nothing of that work myself. In the interval of learning I must surely have a foreman. And I gave myself a promise that I should not degenerate into the farmer who walks round his buildings once a week on Sunday morning to show his guests his animals, his animals his guests.

2

The Valuation

ALTHOUGH I WAS BORN IN Canada I went to school in England, and returned afterwards to my birthplace to work. Many of my school holidays were spent in Wales, and during them I had made the acquaintance of a family of farmers in Merionethshire. I thought of these people now, and went to see them without delay. I shall always remember the visit. I sat by the fire in the huge kitchen, which was a replica of the Dyffryn kitchen and of a thousand others, tucked away in folds of the hills. The three sons of the family sat to tea on a bench along the wall. They were all big and sturdy. So was I for that matter. The father sat at the opposite side of the fire and looked at us four young men.

'It be a recruiting sergeant is want here,' he said. A recruiting sergeant must have been listening, as events turned out, but not of the sort he meant. As I ate my tea of light-cakes, currant bread, and blackberry and apple jam I told my troubles. The bluff old farmer waved them away at once. He refused to see anything incongruous in my unexplained urge to farm, and with a patriarchal gesture offered me as bailiff his son Caradoc. The gift was perhaps not quite so spontaneous as that, for I had heard that Caradoc, who possessed college diplomas as well as practical experience, wished to take a job for a while. Caradoc seemed pleased, and was congratulated by his elder brother, Glyn, and his younger brother, Dai, before these two went out to do the milking. Caradoc and I fixed up terms, and the old man gave me the name of an auctioneer to act for me at the valuation. As I left I think that neither they nor I had any idea how fatefully our paths had crossed.

On my way home I called at the valuer's house. I found him in. He was a short, plump, good-humoured man, like Humpty-dumpty. His name was Bob Rowlands, and he ran a nearby mart. I tried not to let him see my excitement, nor what an amateur I was, but I know now that an agriculturist of any sort is impossible to deceive on his own subject. A scientist, a doctor, an artist, a soldier, a lawyer, can all be impersonated for a short while with some hope of success. A farmer never. More is needed than a glossary of jargon and a studied physical expression. No one can impersonate an earthquake or an acorn, and a farmer is just as much a natural manifestation. Rotund Bob Rowlands creased his shrewd little eyes and, I am sure, read me at a glance. I explained my problem, and was relieved to find that he considered it neither insoluble nor even rare. He offered to act for me, and I told him the time and the place.

I was at Dyffryn early on valuation day. The old man had, with the help of his neighbours, gathered the flock off the mountain the previous day. The sheep were shut in a big stone-walled enclosure above the pens, and as I drove past on my way to the house I felt a tremendous premature thrill of possession. All the bonds and banknotes in the world do not look so prosperous nor so safe as a flock or a herd. Flocks and herds, land and crops, are real. They appeal to instincts which are as old as man. Man became man because he tamed another animal and because he observed a seed sprout. My doubts and my nervousness left me. I felt at last that the inexplicable urge which had driven me to Dyffryn sprang from sound premises. But up at the house my fears returned when I saw the valuer who was set up against my cheerful little champion. He was a tall, lean individual past middle age, with a sunken, cadaverous face, parchment-coloured. His eyes were as hard and expressionless as stones, and he never smiled. He was dressed in near-black homespun, his narrow trousers turned up over black boots. He wore a starched dicky and no tie. His name was Johns. Beside him Bob Rowlands seemed as ineffective as an impudent small boy.

The old farmer entertained us to lunch. I felt like a prisoner eating with his judges. Johns said little. His employer was as moody as I was. Only Bob Rowlands kept up an irrelevant conversation, whose lightness weighted my fears. The meal dragged to an end, and Johns at last spoke a few words to my man. They produced a paper for the old man and me to sign jointly. We had to agree that we should abide by the terms arrived at by the licensed valuers, and that in the event of a dispute between the valuers the matter should be arbitrated by the Ministry of Agriculture. I was now irretrievably committed. I felt more alone than ever before among these strange people, who must be, I thought, chuckling at my ignorance and licking their chops in anticipation of the financial feast to come. For how did I know whether both valuers were not hand-in-glove with each other and with the old farmer? I remembered all the stories I had ever heard of Welsh trickery. I did not know then that they were quite untrue. Whatever price was settled, I must pay, and doubtless the valuers and the vendor would share out their pickings afterwards.

Then cheerful little Rowlands sidled up to me. He was making heavy weather in the rough going towards the pens, and he looked quite out of place on the hill track in his neat blue suit and bowler hat.

'What price do you think you can get them for?' I whispered. Bob cast a glance at Johns, who was loping ahead like a wolf. The rubicund face lost at once its geniality. The humorously wrinkled eyes became calculating slits. The loose lips tightened.

Jekyll became Hyde. Ingenuous Bob looked a match for any man.

'Under thirty,' replied he, from the side of his mouth.

The year was 1931. The slump was well on its way. But even so this figure was cheap, for there was acclimatisation value to consider. A neighbouring farmer had entered his place the year before, and I knew that he had had to pay forty shillings. I expected to take over thirteen hundred sheep, so that every rise or fall of a shilling meant sixty-five pounds to me. I marvel now at the light-hearted way in which I plunged into Dyffryn. My money was all too little for the size of the enterprise, and had the valuation been at a price unfavourable to me I might well have been unable to meet my obligations at the very outset.

We came upon the pens. The pens are set in a remarkable place. They are built at the foot of a miniature cliff which is perhaps sixty feet high. Through the ages Dyffryn weather has crumbled the rock piece by piece, until a couple of acres of ground are piled several layers deep in debris. Some of the pieces are as big as a cottage, few are smaller than a motor-car, and they lie haphazard in utter confusion. At the sides of this chaotic slope are the pens. Above them is a stone-walled enclosure about a hundred yards each way, and tilted against the mountain-side at an angle of forty degrees. The pens are on a level patch at the foot of this enclosure. They are roughly V-shaped, with the small sorting pen at the apex. Chipped on a stone set in one of the walls is the date 1886.

Just now the sheep were loose in the big enclosure. A dozen men were sitting smoking along the walls of the pens. But at our approach some of them jumped down and sent their dogs scurrying up the slope. Instantly the flock became tense and alert. The sheep nearest to the dogs dashed away from them as they came past barking, and merged into the more stolid ewes, who were not so close. I wondered how soon I should have a dog which I should be able to use so confidently. The big flock was harried to a corner of the enclosure, and began to stream through a gate into the pens, while the hysterical dogs, stimulated by one another to the keenest competition, tore to and fro to keep them in bounds. Presently the last ewe took a nervous running jump over the threshold, and the gate was slammed to. The dogs leaped on to the walls, where they raced backward and forward, tumbled stones on to all and sundry, yapped, and fought. As for the men, they transferred their attention from the sheep to me. They were a sharp-featured, dark lot. One or two were short, most were of medium height; all were lean, wiry, and full of nervous energy. They looked alien and unapproachable. I wondered hopelessly how I should ever come in tune with them.

'That's him – in the boots!' cried one excited individual, with a nudge at his neighbour, who looked with interest at my Canadian footgear. It is always uncomfortable for a novice to mix with experts on their job. The amateur feels that he is despised, and from that he becomes still more awkward and useless. Then and there I decided to eliminate all basis of comparison between my expert neighbours and my inexperienced self. I could not hope to bluff these people that I knew anything about farming; better then to be so frankly ignorant, that they would have no misnomers, mistakes, nor mis-directions at which to laugh. And when bit by bit I should come to learn the work, each unexpected piece of knowledge that I showed would startle them to applause.

The old farmer was chewing his pipe with emotion. He joined a group of friends, and every one began talking in Welsh. The two valuers entered the pen and leaned for a little against the wall, while they regarded judicially the many hundreds of milling, bleat-ing sheep. I felt very lonely, and was young enough to feel awestruck at all the commotion for which I was responsible. But though I did not want to pretend to a knowledge I had not got, I certainly did not want my neighbours to think me overcome. I leaned against the

wall and lit a pipe with a great show of unconcern, as if I was for-
ever buying flocks and mountains and was beginning to weary of
the pastime. I think my casual air deceived the audience, for later
I had to struggle for years against the handicap of being thought
a millionaire.

'Come on, sir! You can spare another pound for the heifer.
Money be nothing to you.'

The peaks rose around me, inscrutable of form. Peaks are feminine, wilful, sadistic. They accept homage and break the suitor. To
this day I am uncertain whether they have broken me. They were
conferring together then. A breeze sprang up as they whispered
one to another. They wondered whether I should be given a scul-
lion's place in their kitchen, or whether I should be encouraged
till they had excuse to crush presumption, or whether I should be
ruined before I had so much as paid court to them.

The valuers had begun to wade about among the sheep. This
method of purchase seemed more and more unbusinesslike to me.
How did I know whether my man was competent? Or whether he
was honest? And any price might be put on the nebulous acclimatization value. But Bob Rowlands looked more like a comedian than
a crook, and would only ruin me unintentionally. A dumb show
was being acted in the pens. The vulpine Johns would catch a ewe,
obviously a fine one, and would show her sound teeth, her healthy
eye, her clean hoof, her close wool, her good bone. Little Rowlands
would counter by grabbing a specimen at the other extremity of
condition, with broken teeth, pale eye, loose wool, light bone – all
mountain ewes have sound hoofs at home. Then this spasmodic
energy ceased. The two men drifted to the wall again. Johns lit a
pipe, Rowlands a cigarette. Their conversation seemed to become
generaL They talked of politics, of world affairs, of agriculture, of
sheep farming in particular, and at last of this flock. Now and then
one or the other obviously made an offer, for a hand was held out

for the other to strike in the time-honoured method of reaching a bargain. But each as yet scorned the outstretched hand. And at each failure the expectant crowd of men clustered about the old farmer let out their pent breath in a hiss. As for me, I welcomed each respite, for I was now resigned to immediate ruin, but did not want the certainty confirmed.

And suddenly from the pens came the sound of a smack. The deal was done. The old farmer and his supporters were holding their breath. Some were about to suffocate. Bob Rowlands came bouncing through the sheep towards me, his face creased into a grin. I leaned over the wall and tried not to look importunate.

'I've bought the whole flock for you, including yearlings, at one price,' he said.

'Oh, good,' I answered. 'Er – how much?'

'Twenty-six shillings right through,' said Bob, and his wink betrayed his pleasure.

But the old farmer and Johns were not so happy. The price was lower than any that had been heard of for very many years. As it turned out the value was fair enough, for in 1932 good ewes were being sold for a pound. But none of us fully realised then the drop in price which was to take place. Why then did Johns agree to a figure which was much below that at which he aimed? Because he had been transfixed by the silver spear of eloquence. I have in later days seen Bob Rowlands at a house sale of furniture persuade a cold, wet, reluctant crowd to call individual bids for the back of a hairbrush, bald save for a dozen bristles. I have learned much by watching Bob at work. There is no limit to the capability of the human tongue. A man may wish to sell a beast. He may insist on twenty pounds. A neighbour has offered twenty pounds, and is on his way over to close the deal. Yet an astute buyer arriving in the interval can persuade the vendor to sell for nineteen pounds, and the poor chap will not be able to explain later to his angry neighbour why he did it. But at any rate the work was done. If I had had to pay the previous year's price of about forty shillings, I should have been ruined. The ingoing would have cost me nearly a thousand pounds more.

At once several men came up to congratulate me. One was a fine-looking old man named Robert Owen. He farmed Henblas, on my boundary. He was tall, well built, with grey hair and moustache. He wore oilskin leggings over an old tweed suit, and a stiff dicky without collar or tie. In Wales, land of few surnames, a man is known by his farm. Owen was always called 'Old Bob Henblas'. Then there was Iorwerth Williams of Pen-y-Bont. Iorwerth Pen-y-Bont was the son of an adjacent farm. He was a smart-looking young man, quite emancipated and modern in outlook. Because he was less old-fashioned than many local youths I found it easier at first to establish points of contact with him than with the others. And tiny Davydd Price, Pentref, added his voice. Davydd Pentref was an improvident little chap of middle age, who lived somehow on the scanty proceeds of a few hundred acres of rock and heather. He had four children. His wife saw to it that they were all well fed, clothed and shod. Most of the other men present were shepherds, and would no doubt be called on by their masters to give a detailed account of the proceedings. There was no inquisitiveness in this. After all, big events were rare in the hills. And it was important to them in such a place to know what sort of neighbour they would have.

Meanwhile the two valuers had gone off towards the various sets of buildings. They were to value the hay and the implements. And at the pens the men were lighting a fire. A bucket of pitch was melted over the flames, and one by one the ewes were stamped on the rump with a letter. They stamped 966. I knew now that there were these ewes to pay for, and also three hundred yearlings and some twenty-five rams. It was also explained to me by Bob Henblas, who was friendly from the first, that it was difficult to gather any mountain clean in that district. For instance, a gale from a certain quarter would drive sheep to lee slopes for shelter, for on an open mountain there is nothing to stop them crossing the boundary. And in the broken rough country the wild ewes often dodge the dogs. He said that I might well expect another fifty or more head to come in unstamped from time to time. It would be my responsibility to report these, stamp them, and pay for them.

The lad who was working at Dyffryn and who was staying on came over to me. Did I want a shepherd? If so, his father wished to leave his present job. I found out where he lived, and when the last formalities of that fateful day were completed went to look for him. He was working as shepherd on a nearby mountain farm. The quaint lime-washed house mounted the hillside in a series of steps, for the foundation of each room was on a different level. I did not have long to wait, for presently I saw a man coming off the mountain. He was short and spare. He had a bright face, red-cheeked, and his accent was partly South Wales, which he had acquired when coal-mining during the Great War. But his knowledge of English was good, and that was useful for me. I had heard of him by repute, and was told that he was the most conscientious worker in the district. I liked John Davies on sight, and quite soon he had agreed to come to Dyffryn. That was in 1931. He is at Dyffryn still. He is one of the few inherently honest men I know. And I believe that Dyffryn has long since laid her spell on him.

I nearly made an enemy by engaging John Davies. Davies was at that time looking after the sheepwalk where I found him for his employer, who farmed at Bettws-y-Coed. Thoughtless and ignorant, I engaged Davies without reference to his employer. This was not only ill-mannered, but contrary to custom. I should have asked his master first whether he had made arrangements to replace him. This is, I know, a formality, but my action savoured a little of enticement. By good fortune Davies's employer, my neighbour, was a direct-spoken man, and one too big to take petty offence. He tackled me one day about Davies. I saw that I had done wrong, and he saw that I was willing to learn better. We became friends, and since that day he has repeatedly told me to consult him about my many problems. Much of whatever I have managed to learn has come from him.

3

The Prelude

DYFFRYN HOUSE IS A PLAIN, rectangular building built of heavy, dressed stones and roofed with Caernarvon slate. There are no windows in the end which faces Snowdon, and the whole of that side of the house is slated to withstand the rainstorms constantly flung upon it by the west wind. Downstairs at the back of the house was the big kitchen which I had entered on my first visit, and leading off it were two large dairies. In the front of the house were two good-sized rooms, separated by a narrow hall. From the hall the stairs led up to the bedrooms. There were six. Four smaller ones were over the kitchen and pantries, two large ones over the two front rooms. The plain house had character by virtue of its very unpretentiousness. It possessed strength and utility, and seemed to say, 'Well! Here I am. Take me or leave me. I've no frills, but I know my job.'

And the house did know its job. No one who has not experienced them can believe the strength of Dyffryn gales, nor the destructive force of heavy rain flung at sixty or seventy miles an hour against stone and mortar. I intended to put my men into the two cottages, so that the big kitchen in Dyffryn house would be wasted as such. I decided, therefore, to turn the larger of the two pantries into a kitchen, and to make the kitchen a sitting room. Those were the days before Dyffryn taught me to use my hands.

Now I would cheerfully undertake any alterations or constructions with the help of John Davies or his son Thomas. But at that time I had up a firm of builders from Liverpool. I have since learned to avoid bringing aliens to the place. Dyffryn will not have them. The builders moved the range into the new kitchen without much difficulty. They set it up, fitted a water-heating system, and replaced it in the old kitchen by a dog-grate. But now water had to be introduced into the house. There was a tap in a washhouse near the back door, but the tank on the mountain-side from which its pipe was supplied was not high enough to feed a cistern in the attics of Dyffryn house. It was necessary to sink a tank higher up the mountain and to extend the pipe up to it. We dug the old tank out of its resting place. It was made of slate, grooved and jointed, and clamped together by threaded iron rods. The iron was long since eaten through by the acid damp peat soil, and it crumbled to the touch, so that the six slabs which formed the tank fell apart. The new location for the tank was some way up the hillside, which was just here precipitous. A couple of the builders' men tried to carry one of the smaller slabs to its new position, but speedily gave up. There was deadlock then until John Davies came stumping past. He always walks at an unvarying pace, with his toes out and his knees bent. 'Be it too heavy for you fellows?' asked Davies, grinning at the slab and the men. 'Hup the flamer on my back!'

Davies got down on one knee, and the men lowered the slab gently on to his back. We expected to see him flattened slowly into the earth, but he rose steadily to his feet and plodded on at an even pace up the mountain. This was my first experience of the Celt and the Stone. A Celt will impatiently snatch a boulder from the crow-bars of a gang of Saxon workmen and trundle it rapidly away. He will lift alone enormous stones to make a coping on a drystone wall. He will slide with a touch great flat slabs to bridge a stream. There is no question of strength. The Celt has an understanding with the stone; they have lived together for so long. The stone is taken off the hard-won fields, and is built into the walls which enclose them. It is chipped away to make room for the foundations of a house, and the pieces fit into the walls. It is quarried, and split to slate the roof, and no quarry yet, however up to date, has found machinery to split slates as well as the Celtic hand. The stone is made into gateposts, stiles, drains, floors, shelves. It is at once the servant and the enemy of the Celt. But from their age-long association they understand each other. Davies will go out to split a slab off a fifty-ton boulder, armed with a pickaxe only.

'The grain do run this way,' he points out, and after a few gentle taps the rock disintegrates.

Or he will wish to use a big rock in the foundations of a wall. 'Let me lay hold on the flamer by myself,' he says. 'There do only be one place to catch hold on a rock.'

The English workmen had more trouble when they came to knock a hole in the wall of the former kitchen in order to fit a new window. The men who had built the place had tossed some half-ton pebbles up into the walls. When at last one of these was prised away, daylight showed in one of the front rooms as well. So I had two new windows. The work on the house went very slowly. It was, of course, winter. The men were lodging in the cottages, and, having walked on the level all their lives, had no muscular facilities for mounting the hill to Dyffryn house. This delayed their start each morning. And when they stepped through the back door to do outside work the wind rose and snarled at them, and hurled sheets of water or handfuls of hail at their shrinking bodies. It was spring before the last of the dust and the plaster was swept away. But by then I had an elaborate bathroom, running water in the bedrooms, and the interior downstairs was arranged to my liking. Dyffryn had become an oasis of comfort in the wilderness.

At first my mind was divided at Dyffryn. I knew that my speedy possession of the place had been impetuous, that it had been a case of love at first sight. Dyffryn is so strange a mixture of wife and courtesan: loving, wanton, staid, provocative, calm, furious. My love for the place was at first bewildered by the many caprices, and it was superficial rather than deep, held by whims rather than by grandeur. And I was afraid that life would pass me by. I did not regret my purchase. I was glad that the place was there for me to come back to. To come back to! There was so much I wished to see. I wanted to go again into the stillness of the Canadian woods. And there were strange races of whose countries and customs I had read. There were the priest-ridden Tibetans, perched on the snow-swept roof of the world and the teeming Indians, with their queer philosophies, and the countless patient Chinese, and the industrious, surging warriors of Japan; and colourful European nations, their peasantry a-jingle with dance and music; and slow Russians; and fatalistic Arabs; and black Africans whose legends outrun history. The riddle of life puzzled me, and in my young folly I thought that the answer lay broadcast; that the solution was to be found by search. I know now that the riddle and the answer journey with the traveller, for the riddle and the answer are one.

But the recklessness was on me. My whirlwind courtship was consummated. Dyffryn was mine, to have, to hold, to await me. So away to fresh fields. For youth is a glutton, not an epicure. Perhaps a touch of fear was in my haste. Some qualm of my self-confidence whispered that my conquest was rape, not marriage; that the hills had not accepted me. Their cold, serene heads chilled the frenzy of my first worship. Mentally I began to rant at them. They thought me not good enough, not true enough, not enduring enough. For calm must be reached through storm, and peace through pain. I reacted. They would not let me be of them; then I would try no longer to adapt myself. I would use the hills, use Dyffryn, walk roughshod. And now, to show that I was no mewling lover who crawled to heel at the sign of displeasure, I would away to the Canadian woods.

I had a friend in Canada, perhaps the finest I shall ever have, and we arranged to spend a winter up north. My bailiff, Caradoc, was running Dyffryn. He was efficient and hard-working. It seemed to me that I might well leave him in charge while I tried to cool my hot blood and to steady my whirling mind. I arranged to leave for Canada in September. One warm day during my first August at Dyffryn, Caradoc came past the house. He was very hot. I suggested a swim, and he was delighted. We took two bathing-dresses and drove down in the car to the twin lakes. There is a point on one of the lakes where the opposite shore is only three hundred yards away.

'I'm going to swim across and back, Caradoc,' I said.

'I'll swim one way with you, and walk back round the shore,' he answered.

We slipped into the water and swam lazily, revelling in the vital-izing chill. I was a little way in front of Caradoc at the halfway mark, when he called to me. No thought of disaster had been in my mind, but I shouted back, 'Is anything wrong?'

I needed no answer. Caradoc was flailing wildly at the surface, and was about to sink. I reached him before he went under, and turned on my back, with his body, face upward, resting on mine. I needed both arms to hold him, but my legs were free to kick. I felt very lonely. There was no sound but the gasp of my breath and the plash of water. And beyond the back of Caradoc's head was no view save the gentle blue of the sky. The weight of the body – for Caradoc was a heavy man – often submerged my head, so that I swallowed water and lost precious breath as I choked and coughed. Caradoc lay quite still, though he twitched now and again. My legs began to tire, and I felt a cramp in one calf. I twisted my head to look at the shore. It must have come nearer, but I dared not

45

believe it had. How long we had been struggling I do not know, but several people were on the bank. I turned my head back again, and we continued our slow journey. I tried to count. I promised not to look again before twenty, before ten, then before five. Each time the shore seemed as far away. I was looking now after every kick, and realised that I was in a panic. I refused then to look at all, and hoped our tedious course was straight. My strength was nearly gone. I had no hope of towing Caradoc to the shore. I often wonder what I should have done if the problem had not been solved for me. As it seemed to me in those few seconds, I was certain I could not take him farther, and I was uncertain whether I could get out alone even if I were to abandon him. For the water seemed as thick as glue. My weakened kicks were ineffective. I had taken in a great deal of water. We both sank a little, and perhaps the douche roused Caradoc, for he struggled clear of me and commenced to swim. A weight came off my heart as well as my body. I started alongside him, gasping encouragement. The shore was coming perceptibly closer, and a lorry driver was wading in towards us. And then Caradoc disappeared. There was no struggle, no appeal of upthrust arms. He just slipped quietly below the surface. The lorry driver was swimming out now. I shouted to him, 'Can you dive?'

He shook his head. A wave of anger came over me that Caradoc should be cheated of his life when safety was so near. The anger brought strength, and I like to think that the strength might have come to me when, before Caradoc had begun to swim, I was wondering what to do. I dived after him, and saw a white gleam deep in the clear water. When I came nearer I found Caradoc slipping silently along above the tops of the waving weeds which sprouted from crevices among the boulders on the floor of the lake. I took his arm, and he came with me as easily as if he had had no weight at all. We broke surface, and the lorry driver grasped Caradoc's other arm. In a few moments we were on shore. Several cars had stopped on the road. One man went for brandy, another for a doctor. I worked at artificial respiration until my numbed body sweated, but the form beneath my hands remained inanimate. Caradoc was dead. There was no water in him. He died of heart failure in the lake, and I learned later that he had been warned against swimming.

He was a pleasant, cheerful, honest man, and his memory lives after him. The Merionethshire farm which was his home was plunged in grief, but in a while the youngest son, Dai, offered to take his dead brother's place. The old people seemed to want to send him, and in a short time he came. I put off my departure for Canada until Dai became properly acquainted with Dyffryn. My Canadian friend departed for the north, whither I was to follow him. I sold my car and booked a passage. One morning a cable arrived from Canada. My friend was dead. He had been dragging a shot deer in a canoe over a frozen lake. The ice gave way, and the freezing water speedily claimed its victim.

The months rolled on. My heart was not in my work. The inwardness of the place and its routine was unknown to me. I was not of Dyffryn, nor did I know Dyffryn. The tragedies which had prevented my leaving had shocked me, and in a way I blamed Dyffryn for them, quite unreasonably. Again I made up my mind to go. My friend's body had never been found, but there was a chance that it would be discovered when the ice broke.

I wanted to be there to search for it. Dai had proved himself competent. In February I sold my car and booked my passage to Canada. Two or three days before I was due to leave, Davies and I found Dai lying in the road beside his smashed motor-cycle. His leg had a compound fracture. We rushed him to hospital. I postponed my sailing. I thought that within a few weeks Dai would be out and about with his leg in a plaster cast.

One day the hospital telephoned. Would I go down? A grave

surgeon met me. Gangrene had started in the leg, and it was necessary to operate. They suggested that I should break the news. Dai was lying on his back in bed, his swathed leg propped high. His face was sweating with pain.

'Your leg's taken a turn for the worse, Dai.'

'It hurts a bit.'

'They say there's poison in it.'

'I thought it was slow mending.'

'I'm afraid they must operate, Dai.' 'Operate? Why? Oh! You mean—'

'They say so, Dai.'

Dai lay and gazed out of the window. I can picture his grey face now. Then he turned to me and grimaced what he meant to be a smile.

'They're welcome to the damned leg,' he said. 'I'll be able to get some sleep then.'

There is nothing so fine as courage.

I went back home after Dai's amputation. My mind dwelt and dwelt on that farm kitchen where not so long before I had had tea with him and with his brothers. The recruiting sergeant had been busy enough in the interval. I wondered if the old couple secretly cursed me for crossing their path, for outwardly they made no reproach.

John Davies began to show his capabilities. It fell to me now to direct the farm operations. I realised how little knowledge I had picked up. I advertised for another bailiff, for though Dai was welcome to his job again, he knew that Dyffryn has little mercy on the injured. I put all my energies and all my time to learning the trade I had neglected. And I found that while my eyes had been on the ends of the earth a thousand strange and interesting things were parading unseen by me past Dyffryn door. I did not now say, 'Rewire that fence'; or, 'Gather the sheep for dipping'; or, 'Watch

for that cow to calve'; or, 'Drain that bog'; or, 'Mow that hayfield.'
Now I hammered the staples into the fence-post; took my dogs up
the Glyders to the gathering; heaved on the calf's slippery forelegs
at a difficult birth; sliced the quaking marsh with an old hay-knife,
till the loosened peat could be heaved out of the ditch with a fork;
toiled in the hay from dawn till the mountain dew came down at
night. And I penetrated the indifference of Dyffryn and her hills, till
I touched the warm heart. My men and my neighbours came suddenly into perspective. We spoke all at once a common language,
though there was a trifling difference of tongues. Bit by bit they
taught me. One said, 'Go to marts and learn values. And if you buy
wrong you can only be one bid worse than some other fool.' And
others said, 'Buy hardier rams'; or, 'Lime your hay-meadows'; or,
'Dose the flock for liver-fluke'; or, 'Collect a stray ewe from Llanberis'; or, 'Watch the mare's feet for grease'; or, 'There's a Dyffryn
ewe stuck on a ledge on Tryfan'; or, 'Don't sell your wool too early
this year.' I had a score of bailiffs, a hundred experts, to help me. I
ceased my search for a foreman.

The day I bought Dyffryn there was a change in the old
farmer's attitude to me when I returned from the mountain.
Whether I knew it or not, I had fallen in love with the place during
that wet scramble through the mist. The farmer had known it by
some subtle instinct, and had warmed towards me. And now, as the
manifold interests and excitements of the life began to flood my
consciousness, so that I worked and worried and asked questions
from daybreak till dark, the hillmen, my neighbours, sensed at once
the change of heart. As the spell of Dyffryn strengthened over me,
until my last resistance went and I became her servant, the people
of the district admitted me more and more to their circle.

The Welshman is not simple to understand. He listens to the
jests of visitors about his 'Look you' and his 'Indeed to goodness',
and he hears much from them about his deceitfulness, hypocrisy,

and dishonesty. And when he has the chance no doubt he repays their bad manners by adding a little to the bill. It is his form of retaliation, and many people prefer it to a smack in the eye. There is no country where the precept 'Do unto others' has such point. A Welshman reciprocates your treatment of him. Swindle him, and his tortuous mind will set to work to confound you; help him, and his warm nature insists that he prove his gratitude; mock him, and sooner or later he will contrive to make a fool of you; praise him, and he continues zealous to deserve your good opinion.

And the Welshman's mind is not so direct as the Saxon's. He rarely comes out with a definite question or answer. But he hints at what he wishes to know, and gives dues from which one can learn his replies. This is no reason why he should be accused of duplicity. Why, a woman of any nationality has just the same traits.

So I learned about my neighbours, and came to like them more and more. And as my regard grew, so did their kindness. The splendour of flocks and herds began to obsess me. I started to love my stock as well as did any desert nomad – beginning with Abraham. The points of a bull or a ram, the inherited wisdom of a flock, the miracle of birth and survival, became unending wonders to me.

And it was then that the Gods thought fit for me to meet Esmé. On a mid-winter day I caught my first glimpse of her in a town street. She was small, lithe, with a hard, slim body and the face of an elf. She was as dainty as a Dresden shepherdess. She had the eager, parted lips of a child and the cool grey eyes of a woman. We were engaged just a week later. In the early spring, when the grass was a-rustle with growth, and the trees gave budding promise, and the birds and the beasts preened and stretched in the luxury of the sun, Esmé came back with me to Dyffryn.

And our honeymoon was spent among the springing lambs as they woke one by one to taste the first joys of life. For me the prelude was played. The drama could begin.

4

The Gathering

ONE OF THE MOST TYPICAL functions at Dyffryn is the gathering. Dyffryn mountain is four square miles in area, broken, precipitous and high. Word is sent to our neighbours when we wish to gather, and if our day clashes with no other farm's, twelve or fourteen of us will take our places at the set time. We begin our drive at the short east boundary. The men string out in a line from the lowest point up to the height of land, and sweep the length of the mountain. But not all of us go east. Some scramble up from the Ogwen valley to guard the long ridge, and to turn the sheep over the Dyffryn side of the divide, as they race away before the dogs of the gatherers. And three miles ahead the farms on the western boundary send men up through the huge hollow of Cwmffynnon to act as stops between Glyder Fach and Cwmffynnon lake. For were the sheep unchecked here they would keep on before the dogs till they crossed the pass of Llanberis, to take refuge in the screes of Crib Goch, the buttress of Snowdon. When the line approaches the west boundary it pivots on the bottom man. Those high up wheel round, until they have descended 2,500 feet and are below the mountain wall. And the thickening flock is swept back again for three miles below the wall to the pens by Dyffryn house.

The first gathering of the year is in late March. We have not seen the flock as a whole for months, and we want the ewes down because of the near approach of lambing-time. We lamb our ewes late in the hills. The newcomers do not appear until early April, for we hope, often vainly, that the weather will be kinder then. The weather rules the gatherings too. Sometimes mist and gales prevail for a fortnight or more, and we must wait anxiously for a clear day. And the blustery March gathering is more difficult than most. John Davies is always about Dyffryn house early on these mornings, prowling restlessly around, anxious to be reassured that the mist draped in folds over the curves of the hills will clear, or, if the day is fine, that bad weather is not drifting over from Snowdon. By eight o'clock a group of men assemble by the house. Most of them are cold, sleepy and conscious of having bolted breakfast, for they have done an hour or two's work at their homes before coming three or four miles to me.

The wild strangers of valuation day are among these men. In retrospect I cannot believe that they once seemed alien, mysterious and unapproachable. Old Bob Owen, Henblas, does not come. He is rather old now for strenuous work. But his shepherd is there to help. And there are the others, names to me once, personalities now: Gwylim Roberts, shepherd of Pantffynnon, quiet and well mannered; Iorwerth Williams, the son of Pen-y-Bont, a very eligible young man, modern in outlook, who talks to me as one man of the world to another; Davydd Price, Pentref, small, always in trouble, and whimsical about his affairs; Will Jones, the son of Hafod farm, thin and serious, with the weighty utterance of an old man. They are a fine lot. There is no servility about Gwylim Pantffynnon, nor any of the other hired men. And there is no condescension to a blundering amateur from old-timers like Davy Bach Pentref.

At first I had difficulty with dogs. It is almost impossible to buy a good trained dog for hill work. And in any case there is always better mutual understanding when the dog is bought as a pup, so that he grows up with his master. Hill land would be valueless but for the dogs. It is too high and bleak and rough to carry any stock but sheep, and without dogs the sheep would be uncontrollable. I do not suppose that a couple of hundred men could gather Dyffryn mountain unaided. Thirteen men can manage with dogs. Our mountain dogs are quite different from the trials breed. A sheepdog trial is a circus performance, and it is as unreasonable to expect a trials dog to work in rough country as to expect a retriever to draw a sled. The trials dog is trained for a different purpose. If he makes a mistake a hedge will retrieve his sheep for him, but in the hills the animals would be over in the next valley, lost for the day. And the dash is bred out of the trials dog. He has to be all delicacy and gentleness. The mountain dog must have tremendous vigour, for sheep do not wait for him to approach. They begin to move away as soon as a man appears on the skyline, and the dog must cover half a mile of rough country to overtake and retrieve them before they are lost to him. In the busy summer months the dog is working daily on his own farm or on some neighbour's farm in repayment of help, and good stamina is vital.

Dogs seem to fall into two classes. Some want to range behind a flock, driving them, and the others want to circle and retrieve the flock to the master's foot. A shepherd likes to have one of each type, because he is then well equipped for taking sheep along the road. One dog trots ahead, swaggering like a drum major, the other pads officiously behind the flock.

The working instinct is tremendously strong in sheep-dogs. Puppies will chivy hens into a corner and guard them truculently for hours. When the pup first sees a sheep he whines and quivers with an excitement he does not understand; until suddenly he rushes at it, only to recoil in alarm at the giant's thundering hoofs as it turns to flee. But now and again a dog shows no sheep sense. John Davies once bought himself a bitch out of a very good strain, and she did not seem to realise what a sheep was. If she was with another dog she would snap at her companion as he worked, as though jealous and annoyed with him for performing what she thought pointless evolutions. Davies stopped taking her about, because it is useless to try to train a dog until the instinct comes out. For some reason he kept her until she was two years old, and one day she began to work. All the manoeuvres she had witnessed were at her call, and she took her place on the mountain self-taught.

After many failures and disappointments I bought as a puppy a black-and-white Welsh sheepdog. He had a fine white chest, white feet, and a white tip to his plumed tail. His name was Luck. He has grown to be hard-working, shrewd, shorttempered with most people and animals, an inveterate fighter, and wholly faithful. On the rare occasions when I have been away in the car without taking him too, his listening ears recognise the car's note as I pass the lakes. I see him streaking down the hillside to jump the stone wall into the road. And when I stop to pick him up and he sits beside me he becomes suddenly reserved, ashamed of his display of excitement. A little later I bought another puppy, Mot. Luck was two years old when Mot came, for it does not do to train two young dogs at the same time. One young one is a handful, and two react on each other, so that both become uncontrollable. Most Welsh sheepdogs have a strain of Scottish in them. Luck, with his fine ruff, long coat, plumed tail and chunky quarters, has a strong resemblance to the cross-bred collie. But Mot, who is all black, is a taller dog, bigger, and with a rangy build. He has the loveliest eyes I have ever seen in an animal; they are more liquid than a hare's. Surprisingly, Luck has always been good to him. Mot is Luck's disciple in everything. And both of them – bless their misguided, simple hearts! – live for me. One can learn humility through the blind adoration of a dog. On these gathering days Luck and Mot stalk stiff-legged among a score of strange animals, provoking trouble as they water every coign of vantage, until assuredly one would think they had wrung themselves dry.

At last, driven by the impatience of John Davies, we begin the half-hour's climb to the mountain wall, where we always sit for a while to smoke and gossip. Sometimes on misty days the yolk-coloured sun balances on the shoulder of Siabod opposite. The vapour shrinks, till it lies in streaks which writhe upward, twisting like snakes. And as the air clears we see sprawling at our feet the twin lakes like blots of pale blue ink splashed on a map. To the east the skeins of smoke rise as if pulled by unseen hands from the cottage chimneys in Capel Curig. But fair-weather signs make Davies unhappy and restless, and he urges us to hurry while the going is good. As we continue to climb, a man drops behind at every few hundred yards, and squats, his dogs beside him, while the shrinking group moves upward. Presently Davies and I are alone. A long way off along the ridge white specks move restlessly.

'The men have reached their places yonder,' says Davies, and jumps up on to a boulder. The men at the bottom of our line see his midget figure do a little dance, whose animation is mocked by distance, and his voiceless shout is echoed to them by links in the chain. As Davies and I separate, the men send out their dogs to right and left, and the line moves forward.

The best way to gather sheep is to stand on some high point and to clear the ground as far as can be seen ahead, turning them downhill, so that the next man will come across them as he too makes progress. In the end, chivied from man to man, the sheep will reach the mountain wall, and escape will be impossible for them. That is the theory. But mountain sheep are wild. Sometimes they will turn for neither man nor dog. There was once on Dyffryn an old black ewe. I took her over at the valuation, and she was four years old then. She was a big, rough-coated ewe with one wall-eye, and was quite indomitable. Time and again from my gathering position I would see her charge past the baffled dogs, and go bounding down the precipice beyond the boundary. If an indignant dog found his way after her she would rocket up on to Tryfan, to find sure sanctuary in the stony maze. I sold the old ewe at my first September sale, and she went to a buyer who lived twelve miles away. But next March gathering the familiar figure was home again, tumbling wildly away down the screes, unmistakable in her vitality. I thought she had come home to die, and sent her owner a cheque in refund. We saw no more of the black ewe for two years, until at a summer gathering she bustled in before the dogs, burlier than ever with two clips of wool on her back. She was followed by a black yearling lamb and by a white lamb a few weeks old. Neither of the lambs had been touched by human hand. They were not earmarked.

The old ewe died that winter somewhere among the snow and rocks of her favourite Tryfan. Her black lamb carries on her tradition. She is a matron of six today, and is not for sale.

As the top men of the line, Davies and I hurry, for we have much farther to go than the bottom men. Esmé sometimes comes with me gathering, whenever we have a maid in the house to prepare food against the descent of the hungry gatherers. But unfortunately few girls care to live in the loneliness of Dyffryn. And when Esmé does find time to come she is hard put to it to keep up with us. For often when a gathering is half done the mist rolls over so that we lose sight of sheep, dogs and each other, and are forced to abandon our drive and make our way down as best we can. So we always hurry along the Glyders heart in mouth, for fear our whistling and shouting should arouse the sleeping wrath of the gods.

The men from the Ogwen valley meet us under the hump of Glyder Fach. Sometimes we find them huddled under a rock out of rain and wind; in summer roses are stuck in the peaks of their caps as if they were Afghan dandies. The incongruity of a rose on the harsh mountain always strikes me afresh. These men take our places as John Davies and I clatter in our nailed boots up the stone pile of the Glyder. The boundary here is sheer. There is no place where a sheep can escape us. And across empty space Tryfan rises in unbuttressed isolation. Tryfan, the queen of mountains, hard ruler of them all, has many suitors, but none can claim her love. She guards her cold honour with veils of mist, with smooth black slabs of rock, with poised boulders, and with enticing green gullies which bend ever steeper, till the lover slides away in a loose, wet avalanche of stones and moss. Tryfan sacrifices many bloody victims on her cold stone altar, and remains the vestal who offers nothing but a dead embrace.

Near the summit of Glyder Fach Davies veers towards the lip of the precipice as he hurries to start the turning movement of the long line. Somewhere ahead of him will be the men from the Snowdon farms ready to act as stops. I have time to light a pipe here. When Esmé is with me we sit in silence. The place is high, free and glorious. The dogs stay beside us, sitting bolt upright, prick-eared and tense. There are hares up here. Usually they show us only a distant glimpse. Once as we sat there a hare appeared quite close. He was unaware of us, and tripped quietly past, very trim in his snow-white winter coat. He hopped on to a rock and rose up on his haunches, his forepaws held as if they were tucked into the armholes of his waistcoat. His nose wrinkled quite distinctly as he sniffed up-wind. Then Luck stood up, and the hare became a white streak galloping between the boulders, his ears laid flat along his back.

But it was not men nor dogs which had disturbed the hare. From above us came a tremendous hullabaloo. Rapidly approaching from the crest of the Glyder we heard breathless barks, shrill with excitement, and from somewhere beyond came the faint angry yells of John Davies. Luck and Mot shot off, taking no notice of my shouts, and through the rocks streaked a lithe red body. Dogs become uncontrollable at the sight of a fox. They nearly kill themselves in their vain, clumsy efforts to catch up with him, and I do not believe they would know what to do if they did catch him. The brute passed close to us, moving quite effortlessly, as his huge brush balanced him on the turns, so that he glided as gracefully as a slalom skier. The dogs were panting a long way behind. They scrabbled wildly up tilted slabs, dropped heavily off them with a grunt, missed their footing, and somersaulted. The fox, unhurried, disappeared. Bett and Jim crawled back up the Glyder towards John Davies; Luck and Mot returned shamefaced to me.

The sheep who keep the highest boundary are cunning as well as hardy. They take advantage of every diversion to attempt to slip through our attenuated line. Often as I sit quietly here, a string of sheep disturbed by John Davies appear above me, tip-toeing silently back, unaware of my presence. I dare not send Luck after them direct, in case they should break into a run and be lost to him, for they could travel at least as fast as a dog over the rocks. So I motion him quietly behind me. He always eyes the sheep for a moment, then he takes a quick look at the layout of the ground before he streaks away like a flash, belly to earth, tail tucked in, quick to catch my wish that he should be inconspicuous. He makes a wide détour before he turns in to meet the sheep. Round a boulder he suddenly confronts the leader face to face. Both stop dead. Luck will lift a forefoot and advance one delicate step. The old ewe gives a stamp and blows defiance through her nostrils. Luck crouches and creeps one pace nearer.

At that the leader of the little flock will lose her nerve, break and run, her scattered troop careering pell-mell after her. As they pass in front of me headed downhill, Luck swings away and comes back to my feet. He disregards Mot's envy with elaborate unconcern. Mot is much too young and impetuous for such crafty work.

At last comes a scrabble of boot-nails on stone, far away, but quite audible on the still air. Davies has rounded the peak, and can see ahead of him the men from the Snowdon farms.

The way now lies down a precipitous hillside. Strange turrets of rock guard the brink, and beneath them falls away a square mile of country littered like a giant's battleground. This is the great hollow of Cwmffynnon. Slabs and boulders which range in size from a football to a house lie piled in unutterable confusion two deep or twenty deep. From every crack and crevice spring tufts of whin, heather and bracken, concealing the pitfalls.

It was here in late autumn that a German, alone, slipped and broke his leg. From where he lay he could see the climbers' hotel where he was a guest two thousand five hundred feet below him. He had told no one where he was going, and across the valley men were searching Siabod. The injured man made a rough splint from his walking stick, and began to drag himself down over the confused heaps of stone. Towards night the mist crept down. With the mist came rain, and the man lay out that night. On the second day the mist was still there, and the rain fell steadily. All day he crawled, in pain and hungry, unable to be sure even that all his efforts were not taking him away from help, for he could see only a few yards ahead through the clouds. The second night he lay out. He was so close to the hotel that he could hear a dog barking in the sheds. They found him outside the back door on the evening of the third day. After a drink he refused assistance and pulled himself upstairs to bed.

It was here too, a few years after the Great War, that a shepherd stumbled over the skeleton of a man. He had been lying there for about a year, and the crows do not leave flesh on bones for long. No one ever discovered who he was or what he had been doing, and the few shreds of cloth which flapped among the whins gave no clue. A few weeks after this discovery two Oxford undergraduates were hurrying down the mountain on the other Dyffryn boundary, three miles away. Lying in a hollow beside a stream was the body of a woman. She too had been there about a year. The young men reported their find and its whereabouts and returned to Oxford. But the police and their helpers could not find the place, and they had forgotten to take the names of the undergraduates. However, an advertisement in the papers brought one of the youths back, and soon the remains were carried down. A farmer has shown me where the woman had lain just above the mountain wall. He said that no grass would ever grow on the spot again. None has.

On the mountain at dusk, spirits float down to settle on the tops. I often wonder who they are, and whether the man and the woman are among them. The two died together and apart. Perhaps they knew each other, and perhaps they died unaware. I used to think their ends a grim tragedy, but now I am not sure. They were tired, and lay down to sleep in the lap which bore their race. In the brooding peace they were undisturbed by shrouds and tears, by the squeak of a screw in an oak lid and the thump of earth on a coffin. But man cannot let well alone, and the quiet husks were collected in a box and buried amid the odour of decay.

The only tragedy of their end was that it was unremarked by any. Even a sinking stone leaves a fading ripple on a pond.

Luck and Mot and I scramble down this place. Even the dogs find the going difficult, and they obey me reluctantly when I send them out after sheep. Sometimes a pungent, musky smell floats up to me, and soon I see a string of goats on a ledge, ranged like a rank of soldiers. I have counted as many as thirty-five at a time. The dogs become wildly excited, but I manage to keep them at heel, and we pass on. In the dim past these goats used to have ownership, and they were earmarked like sheep. They were turned up into the hills so that they would follow their instinct and jump down to graze dangerous ledges, and so make the places unattractive to sheep. For a goat is more agile than a sheep, and can leap off ledges where the sheep would lie marooned until starvation overtook it. Forbidden fruit attracts animals as well as men, and to eat it leads as frequently to disaster. But now all ownership has really lapsed, though I suppose a farmer can lay claim to any goats on his land. Occasionally a kid is killed and eaten. The flesh is extremely good roasted, and has a strong flavour which is a cross between wether mutton and venison.

In hard weather the poor brutes come down low into the valleys, their ribs sticking out through their skin, and in March the pathetic kids are found curled up in some cranny, while the mother ranges about for a bite of food to keep milk in her udder. The kids are hard to help, for the nanny often deserts them if they are handled, and in any case they do not take so readily as lambs do to a rubber teat,

As we drop down the hillside the intervals between the men become less, and it is soon possible to shout to each other. There is always news to tell. The day Esmé and I saw the fox, Davies yelled that he had thrown his stick at him and hit him on the rump. I shouted back that I had put a match to his tail and sent him off ablaze like one of Samson's vixens.

When I was first at Dyffryn I used to be very worried at the gathering. The neighbours were strange to me, and I had not then realised the area of the place. It would seem to me that I had to walk miles to see a dozen sheep, and I used to suspect wholesale sheep stealing, or I wondered whether mortal disease had not swept through the flock. Of course, there is very little sheep stealing. Now and again some smallholder will not be very particular to return a stray ewe, or there are rare cases of unscrupulous butchers who drive their vans along lonely mountain roads to catch a few sheep with a dog and take them off to slaughter.

The last big-scale sheep stealing near us took place shortly after the Great War. And though large numbers of sheep were being taken, it was only by accident that the thefts were discovered. It is very difficult for a hill farmer to tell whether he is losing sheep, because he can never be certain how many head he possesses. It is impossible to be sure of gathering in the entire flock at one attempt, for sheep double back through the broken ground and are unseen by the gatherers, or perhaps the weather has driven many to shelter over the boundary. Again, a neighbour may have his stock temporarily off his mountain, so that tempting grass is left untenanted.

There is only one satisfactory way to count a flock, and that is to count each sheep as it is sheared. As unshorn sheep come in from time to time after the shearing day they are clipped and added to the tally. But unfortunately few farmers like to take a count, because they think it unlucky. The legend of bad luck has arisen simply because a count never comes up to expectations. Ostrich-like, the farmer refuses to admit the many dangers to which his flock is exposed. Each generation of ewes spends four years on the mountain before it is sold, and during that period fifteen in a hundred may die. While the lamb is young she is a prey to foxes, carrion crows and ravens. And when following her mother she falls into streams and tumbles down holes among the rocks. After she is full-grown she may be driven carelessly by dogs till she plunges over a cliff. She is exposed to the dangers of lambing. Now and again a ewe strays away and does not return. Thus when the large-scale sheep stealing was in progress nobody was suspicious. To make discovery even more difficult the thief was operating on one of the few common mountains in North Wales. These common mountains each run with a large group of lowland farms, and every tenant of the farms has the right to run so many sheep in proportion to his lowland acreage. So even if someone had caught a distant view of the sheep-stealer on the mountains he would only have thought that he was a neighbour shepherding. Meanwhile the sheep were being shipped to Ireland from Holyhead, where Irish buyers bid for them in good faith.

But one day there was an important sheepdog trials at Dublin, and some Welsh shepherds saved up for the outing and went over to watch. A shepherd can spot an ear-mark at a distance where a townsman would be searching for the sheep, and it was not long before one of the men saw that the sheep which the dogs were working belonged to him. As the sheep were turned out in threes for each dog in turn, man after man recognised animals which should have been grazing on the other side of St George's Channel. The excited Welshmen rushed behind the scenes and stormed the sheep pens, where they found a representative flock from their common mountain.

The thief was sentenced to eight years' imprisonment. He left behind a young son. The boy was cared for by relatives, and is today a barrister. But there is no moral to this story.

I soon learned to discount the possibility of sheep stealing. As the line of gatherers neared the end of the beat I would see group after group of ewes melt together like blobs of mercury. Twos would become dozens, and dozens scores, until long white strings were winding down at a trot through the rocks, harassed by dogs and men, and the valley would seem full of bleating animals.

Flocks and herds appeal to the atavism in us more even than do fields of rippling grain. Because the first farmer, a hairy, apelike creature, certainly tamed stock before he learned to cultivate. And a seething flock or a bobbing herd is a brave sight. As we turn the sheep off the side of Cwmffynnon through the mountain wall, so that they spread for a quarter of a mile over the lower land till they strike the road fence, cars, cyclists and walkers stop. Who knows what strange stirrings of atrophied heredity these holidaying city folk feel? There is more than curiosity in their excitement. They find their escape from the false values of Frankenstein civilization in the sooty soil of allotments, in rose trees among the ash-cans, in

window-boxes. But these artificialities are drugs. And here is the
real thing! Here pours a vast flock, in high summer to the tune of
thousands of head, in primitive splendour, in an untouched setting.
The spectators are like children whose model engine is replaced by
the Flying Scot. We are three miles from the house and the pens
when we cross the mountain wall, but no vagary of the weather
can rob us now of our sheep. Presently the panting dogs escort the
rest of the flock into the park below the pens, and we tired men
plod up the hill to see if Esmé has food ready for us.

5

The Lambing

WHEN THE FLOCK IS SAFELY DOWN after the March gathering we pen it, being careful not to jostle the pregnant ewes. We examine the mouth of every sheep, and separate the two-year-old sheep from the three- and four-year-olds. There are no yearlings as yet on the mountain, for they are not home from wintering.

A sheep's age is told by the teeth. Up to eighteen months the mouth has the tiny lamb's teeth; then the little front ones are forced out by two broad, permanent incisors. At two and a half years these incisors are flanked by two more, and in each subsequent year by two more, until there are eight broad teeth in the upper jaw, and the sheep is known as full-mouthed. After that, like a nine-year-old horse, the ewe is aged, and one guess is as good as another.

The two-year-olds are taken to the bottom of the valley and put in the long meadows between the road and the river, where the pasture is good, and where stone walls and buildings give shelter. These ewes are about to bring their first lamb, for as yearlings they had wintered away and were kept barren, so it is among them that we expect to find most of our lambing troubles. Along the bottom land it is easier to keep an eye on them. Davies wheels out hay-racks, which he replenishes daily, but few of the ewes will eat. A mountain sheep does not take kindly to artificial foods. One year about Christmas-time two wether lambs worked their way down from the hills. They were so thin that it was easy to lift them one in either hand. They had become too weak to forage for themselves, and we put them in a stackyard with a rack of hay and a pan of meal. They refused the hay and trampled in the meal. We poured gruel down their throats, but it only delayed the end. They starved to death.

The older sheep are turned into a six-hundred-acre enclosure below the mountain wall. The enclosure is called the ffridd, and is pronounced freeth. Every hill farm in Wales has its ffridd. It corresponds perhaps to an English paddock, which I always think of as a handy, fenced-in place near a house.

As soon as the flock has reached the scene of its accouchement Davies begins to patrol the two-year-olds, his son Thomas the ffridd, and I both places, to look for early lambs. Between times Davies begins to construct what he calls 'corners'. These corners are just little temporary pens made of a gate propped in the angle of a wall or of an odd length of wire-netting. In them he will presently imprison sick ewes, ewes who will not suckle their lambs, or who misbehave in any other way.

He calls this action 'bringing the flamer to her senses'.

I am busy making sure that I have all the medical supplies that may be needed. We are going to want disinfectant and carbolic soap to clean the hands before we help a difficult birth, and antiseptic pessaries to insert after a lamb is pulled. If septicaemia sets in after these precautions I have a hypodermic syringe with which I can inject a vaccine near the anus, and this has before now brought back the patient almost from the dead. For the lambs we need thermos-flasks to carry warm milk, and bottles with rubber teats for feeding.

A hundred firms sell a hundred veterinary products. The more modest are for specific cases; others embrace all animals and all ailments in a generous, sticky grip. There are Red Drinks and Black Drinks and drinks of every hue; and Cordials, Painkillers,

Appetisers and Revivers. Some are good, some are bad; fortunately most are harmless. But common sense and two penn'orth of Epsom salts will often do as well as the best of them.

Occasional lambs are arriving by the end of March, the result of an unauthorised early visit to the mountain the previous November by some stray ram. But on the first day of April the patrols have to be interrupted while we go to fetch home the yearlings, whose wintering time is up. They are scattered among half a dozen farms fifteen miles away in the warm Conway valley. By persuasion I arrange for each farmer to bring his lot to a central point part way to Dyffryn. While I wait to count each successive flock as it is brought to the meeting place Davies and Thomas wander off with the early arrivals. When they return they come from the direction of the inn. They are quite unused to beer, and set off on their nine miles' walk behind the yearlings in a most cheerful frame of mind.

Then it is my turn to receive hospitality at the inn. But my drinks cost me two hundred pounds as I pay cheque after cheque for wintering. In turn each farmer beckons me outside, and hoarsely whispers behind a heavy hand, 'I'm not saying nothing against nobody, mister, but you could see I done the yearlings better than none of them.'

Before dark that night the footsore sheep have reached home and are turned up on to the mountain, until now empty of sheep. All summer the yearlings will stand out among the other ewes, when those are eventually returned to the mountain after lambing, because of their colour. When a sheep is feeding on a flush of rich grass pigmentation takes place in the wool, which becomes yellowish. Farmers are remarkable people. When they buy a dip they purchase one which has a 'bloom' mixed in. The sheep scramble out of the dipping bath a rich yellow ochre. Everyone knows this healthy colour is false, but the buyer still gives more because of the cosmetic.

It does not pay to economise on wintering. A poor farm will charge seven shillings a head, and a good one only two or three shillings more. But at the end of the period the difference in the value of the stock from the respective farms will be much more than three shillings. The yearlings from the poorer places will be stunted and unthrifty, and some will have died during the winter. They will in turn produce poor lambs, which will perpetuate their disabilities until rigorous culling and vigorous rams have raised the standard of the flock again.

The yearlings from the better farms will be big-framed, with ample room for growth. They will have good bone, and before autumn they will be strapping ewes well able to resist the winter.

And they will give little trouble at their first lambing the next spring.

After April comes, Esmé and I can look down from the house each morning and see more and more white dots on the green grass of the meadows. And up behind the house in the ffridd the lambs show clearly against the bracken.

Some days the sun shines, and the still valley becomes a-rustle with growth as the brisk spring air lures out buds and tender grasses. The Glyders and Siabod sprawl on their sides, like basking giant-esses, who benignly regard our antics in the valley between them. And stretched across the head of the vale, like a hostess reclining on a high couch at a Roman feast, lies the range of Snowdon. Why must we call her Snowdon? Why not the real name, Eryri, the Place of Eagles? To one side of her peak the precipice of Lliwedd falls sheer a thousand feet seamed with gullies, each one etched sharp in light and shade by the slanting sun. No robe ever hung so gracefully from patrician Roman.

Housework goes by the board at lambing-time. Esmé must join me in my travels to watch the fascination of new life. I take Luck with me, because he is responsible and will trot quietly to heel. Mot, too young and impetuous, stares sulkily through a window after us. Are Esmé and I petty-minded? I do not know. At any rate each walk is filled with interest for us and excitement. Now and again we will find twins behind some wind break. Once the newcomers were from a ewe with black face and legs, a throwback to some Scottish strain imported into the flock heaven knows by whom or when, which still crops up once in a thousand births. One lamb was coal-black, the other snow-white. The Welsh ewe rarely has twins. It is quite as much as she can do to feed herself through the winter while carrying one lamb, and to find milk for it after its birth. Or perhaps we come to a hollow, a sheltered place surrounded by a cirque of rocks. At dusk last night there were six sheep there. Today there may be eleven, and in a few minutes there will be twelve, for the remaining in-lamb ewe is close to her time. We scramble quietly up to the top of the rocks and sit down in the sun. I light a pipe, while Luck sits erect, staring at some mid-air vision.

The ewe lies down, rolls on her side, and pants a little. She gets up, grazes half heartedly, and lies again. I take my eyes off her for just a moment, and when I look back she is on her feet once more as she stares in superstitious amazement at the slimy morsel of life which has so mysteriously appeared. The lamb suddenly lifts his head and tries to erect his drooping ears. His startled mother, whose first lamb it is, jumps quickly backward, then, drawn by a power many thousands of years older than she, approaches again cautiously, a step at a time, ready to shy at the slightest alarm. She sniffs the warm object, and begins to lick it clean, and as she cleans, the lamb wriggles, until he is sitting a-sprawl, his legs at impossible angles, as if pinned to him by a blind man.

We get up to go, and as we stand, the mother will come between us and her offspring, futile and gallant, stamping her forefeet, her head thrust out truculently. Luck affects not to notice. A dog is at a great loss when a ewe calls his bluff, because there is nothing he can do about it except bite, and that earns disqualification. We move away, and when I glance back in a little while the lamb is on his feet, his blood-red navel cord trailing on the ground as he stabs his nose blindly at where he supposes food to be.

Esmé and I can never walk far along the bottom land without encountering John Davies. On these warm fine days he is always smiling as he puffs at his pipe, and we can tell at a glance that all is well. Davies has the real hillman's walk, a deliberate putting down of one foot before the other, knees bent a little from continual uphill or downhill slopes, toes turned out, pace unvaried whether he is on the level or negotiating a sixty-degree scree, whether he has just come out or whether he has been on his feet all day. He will sink down on one knee beside us, the nearest he can come to relaxation, and nod up at the ffridd, fair in the sunshine. A tiny black figure is showing for a few seconds on the skyline before vanishing. It is his son doing the rounds.

'Thomas do be having no trouble in the ffridd for sure,' says Davies. He can scent trouble at any distance.

We sit there and talk of many things, while all around us new life is poured into our world with a liberal hand. Davies says that the little weak ewe who always grazes under Clogwyn Ddu is close to lambing, and that he is going back to see her after his lunch, because he thinks she will need help. During the night that big ewe with the speckled face and the long tail, which I had said was the type we wanted to aim at breeding, has had a magnificent ram lamb, marked just like his mother.

I suggest we put a mark on him so as to remember not to castrate him, he will do for a ram.

'You needn't mark him,' says Davies. 'Us'll know that flamer anywhere.'

Later on Davies will pick him out of hundreds more as they struggle in the pens, and we will let him keep his bold manhood.

In a while Davies will say, 'My belly do tell me it be twelve o'clock.'

And we return, he to his cottage, Esmé and I to climb up to the house. On the way we are sure to meet young Thomas, who also has an alarum clock in his stomach.

'Well, it don't do to grumble,' he says, in answer to my question. 'There be a dose of lovely lambs along the breast there have come in the night. That old ewe me and you went to fetch back in the car from Llanberis have had a black lamb. I didn't see nothing wrong nowhere. It do be a treat like this.'

But the idyll fades as a placid image in a pool is shattered by a swelling storm. Our rainfall at Dyffryn is two hundred inches in a year. At Bettws-y-Coed, six miles away, it is forty-eight inches.

Some mornings we wake to the barbaric music of a gale that lashes at the squat stone house until the bed shakes. It is an effort to open the back door against the thrust and vacuum of the wind, and the rain blinds the eyes with stinging blows, so that one walks bent double, leaning against the storm, tottering like an old man. Luck's long coat is blown to each side of him. A parting is made along his spine, and the hairs are plastered to his sides in an instant by the deluge of water.

On these stormy days I go down to the low ground first, to see whether the sheep are clear of the flooded river. The ewes lie behind every hillock and rock and bit of wall. Their long, broad tails, left uncut for just such weather, are curled against the storm as some sort of protection. The mothers have their lambs crouched up under their chests, and wait in stoic patience for the ordeal to end.

I may stumble on a dead lamb, born quite recently. He could never have risen to his feet, and must have been at once flattened into the soaking earth by the gale, for his mother has cleaned the upper side of him, but when I turn him over the under-side is untouched. The ewe hovers near by, the afterbirth just breaking free. It is no time for delicacy, and I send Luck for her. He ranges round her, just a neck ahead, and turns her in tightening circles, until I am able to spring and grab her. I put her on my back, and carry her with her dead lamb to a building.

As we stumble into shelter the diminution of noise and violence is abrupt, and one hears the storm at a little distance as it swoops and screeches round the shed, like a harpy baulked of her prey. Suddenly a figure darkens the doorway. It will be John Davies. He is wearing oilskins over an old overcoat. He has oilskin leggings over knee-high rubber boots, and on his head is a sou'wester over an old felt hat. He is wet to the skin.

From under his coat he drags a puling lamb, an uncleaned, newborn morsel, in which the spark of life, still vigorous, would speedily have been quenched outside.

'His mother be not worth potching with,' explains Davies. 'She haven't no milk for him.'

I turn over the ewe whose dead lamb I have brought in with me. Her swollen, unsucked udder is full of milk.

'She'll make him a proper mother,' I tell Davies, and draw a little of the thick first milk for him to see.

'Wait till I do make the little flamer a raincoat,' says Davies. And he will pull out his knife and skin the mother's dead lamb. We fit the skin over the live lamb and put him near the bereaved ewe, who backs away suspiciously at first; but she speedily smells her substance on his new coat, and trots back, whickering, to make much of him.

I hold the ewe for a minute or two, while Davies puts the lamb to a teat. As the lamb sucks he strengthens quite visibly. He rises on to his back legs, waggles his tail, then rises to his front legs, unsteady but triumphant, and pushes with ever-increasing emphasis at the udder. It will not be more than a day or two before the adoption is complete and the skin can be removed.

Sometimes we come across a ewe whose lamb has disappeared. Perhaps he has been swept away by a stream, or has tumbled into a hole, or has been carried away by a fox. But we can now and then manage an adoption without the skin of the vanished offspring by rubbing the new lamb under the ewe's tail, so that her own smell is on him.

John Davies and I leave our two protégés in the warm shed to become better acquainted. We pause a moment on the threshold, like divers taking breath, before we plunge again into the storm. Outside we yell unheard instructions to one another and part. Disasters never come singly. Before I have gone far I may see a white object quivering in the wind. It will be a lamb's tail. And near by I will find two headless, tailless bodies, with two heads lying a few yards off. A fox kills wantonly. Two bodies lie at my feet, but more may have been carried away. I do not know why the fox cuts off the heads of his victims. It is as if he were a braggart criminal leaving his fingerprints.

Two ewes hover restlessly in the shelter of a wall quite near, whose bloody tails proclaim them recent mothers, and whose distracted manner shows that they are bereaved. I pick up the dead lambs, and with Luck's help drive the ewes into an enclosure. With the aid of the skins they will be persuaded to take a twin apiece. It is quite enough for a Welsh ewe to rear one lamb, so we always try to separate twins.

If the rain drives down all day, by nightfall every building will be holding couples. Many of the lambs will be too weak to trust outside, and many of the mothers will have been so buffeted by the gale that they will not feed their offspring. Lambs will lie in the cottages and in Dyffryn house. These are the really bad cases which have not the strength to suck at a ewe. Some are brought in chilled right through by the cold rain, and are revived in a basin

of warm water and given a little gin. Others are fed drop by drop
with a fountain-pen filler. There may be one that has been almost
drained of blood by a weasel, his body so flaccid that when picked
up he drapes over the hand like a white cloth. The gale may last a
fortnight. If it is prolonged the work gets out of hand, and we have
to leave the old-stagers in the ffridd to look after themselves while
we concentrate on the bottom land. In the buildings the family dis-
tinctions become impossibly complicated. This skin belongs to that
ewe's dead lamb. This lamb has been deserted. Is that lone ewe
with milk in her udder his mother? Or is her son represented by
this skin? Then Davies says, 'That flamer have come to her senses.
I be going to let her out with her lamb into the stackyard.'

And when two or three indifferent couples are under observation in the stackyard, wrongly paired, the confusion becomes worse
confounded. And we dare not keep a ewe in for long, or her milk
will dry up, for it is almost impossible to persuade her to eat artificial food. Yet if the couple are turned out prematurely the ewe may
desert her lamb.

John Davies and Thomas develop an open hostility to Providence. They curse the careless mountains. They are working eighteen hours a day. Their clothes are used up faster than they can be
dried, and they are always wet and cold. But they do not grumble
at the work. They grumble about Dyffryn's tragic trouble.

But, as in Genesis, there comes a day when the floods subside
and the quaking ground becomes firm again. For many days the
mists have veiled the hills, and now the sun woos the filmy clouds
and lifts them to his embrace, so that they die in the warmth of his
kiss. And the mountains seem to shake their streaming flanks like
monsters rising from the deep, and they lie a-sprawl as they dry
visibly in the sunshine.

The ewes come forth from every sort of sheltering place and
begin to graze again phlegmatically, some with lambs at foot who

wrinkle their noses at the world which has been so hostile to them, others heavy, their ordeal still to come. John Davies and his son begin to smile, and daily John Davies comes to me to say some such thing as, 'That ewe in the pen by the top building have come to her senses. I have let go the flamer.'

We begin to look through ffridd, and find that the wily older sheep show little sign of the hard times they have been through. Here and there a lamb is wedged between two boulders in a stream, drowned while following his mother. Where we can, we put slate slabs as bridges, but these have to be exactly on the sheep tracks, or they are not used. In one place is a lamb, dead now, whose living eyes have been plucked by crows. A carrion crow will feed on a lamb but partly born.

But soon we forget the losses of the storm. Each day new lambs come, forty or fifty of them. The percentage of deaths shrinks. Soon the whole valley seems filled with contented couples. The lambs become stronger, and the ewes leave them asleep while they graze farther and farther afield, always to return unerringly to where their lambs lie. And now the lambs themselves are show-ing initiative. Groups of them dispute the sovereignty of a hillock. They chase one another, losing themselves and bleating pitifully, until their mother recognises their voice and calls back as she trots in search.

No miser with his hoarded gold can feel the pleasure which comes to Esmé and me as we stroll now through our doubled flock. The farm has given birth, and we identify ourselves with its labour. We have shared many of the pangs. There is no greed in our eyes as we survey the teeming land; rather are we humbled before the courage, persistence and simplicity of nature. We have been privi-leged, for the gods have performed before us the great play of life. A gigantic orgasm has been succeeded by a gigantic birth. The fluid of fertility has been poured over our hills and valleys by firm hands whose generosity shames humanity. We feel that we have assisted at a miracle. And, of course, we have.

6

The Year the Snow Came

ONE SATURDAY NIGHT IN EARLY March in 1937 the snow came. We often have snow in winter at Dyffryn, but it is seldom very deep, nor does it last very long. And in some way the sheep smell the threatened blizzard, and from our places in the bottom of the valley we can look up at the darkening hills and see them winding down in long white strings, so that they are soon thick along the lower land. When the snow puts down itsbarrage on the mountain the victims have already fled.

But this year the snow came with no warning. The most weather-wise sheep were caught in the early flurries as they made their way down to the mountain wall. There are gutters at frequent intervals in the wall, and in winter these are left open for just such occasions as this, so that the sheep can slip through and descend still farther. But that night the ewes, heavy in lamb, ploughed their way to the wall only to find the gutters buried deep, and as the animals tried to range to left or right in search of an open hole, the screaming wind hurled the snow in masses against the wall and buried the refugees where they stood.

Esmé and I had been away for a few weeks on holiday. In the southern papers there was no hint of the severe weather, and one Monday morning we telegraphed that we were coming home for a couple of nights to see that everything was all right. It took us a day to cover the last thirty miles. After many attempts we reached Bettws-y-Coed at midday on the Tuesday. In answer to our inquiries no one could give us news of Capel Curig. The snow had come without warning during the Saturday night. The roads were lost under drifts, the telephone wires were down over several counties. We knew about the wires. I had lifted the tangled masses time and again while Esmé drove the car under them. We abandoned the car at Bettws-y-Coed, and reached our valley in the cold half-light of evening. The frozen lakes lay dead in a shrouded world. Our lost domain was silent as the grave. The virgin drifts covered the buried road, and nowhere was there a spark of life. We ploughed up the valley, waist-deep in the soft snow. When at last we reached the cottages we were received with exclamations of surprise. Davies had seen no one but our own people since Saturday. Farther up the road were cars abandoned, but where the travellers were he did not know. On the Sunday, he said, it had taken him nearly the whole morning to carry water to the building, a furlong away, where some of the cattle were. It was quite impossible to get the beasts out to drink at the stream in the usual way, for the cow-house doors were buried and the stream was lost.

Davies said that he and Thomas had begun to search along the walls and along places where sheltering sheep might have stood until drifted over. They had found about fifty sheep, which they had carried through the drifts and had set down upon occasional exposed patches of ground which the wind had cleared of snow. There was no food on these places but a little sparse grass, for it was still winter in the hills, and, anyway, several sheep on a patch just a few yards square would speedily kill what herbage there was as their sharp hoofs ground against the iron-hard earth. These patches, too, were there only by virtue of their exposed positions, and the sheep were lying open to the bitter wind.

I asked Davies where the flock was. Year after year I had sold fewer ewes than I had ewe lambs to replace them, and the breed-ing flock had grown from under a thousand to thirteen hundred. The yearlings, fortunately, were still out at wintering in the Conway valley.

For once Davies was at a loss.

'Well, damn,' he said, 'I don't know where the flamers be. I never seen nothing like this before.'

It was late in the day, and Esmé and I struggled to the house to find damp bedding, which we tried to dry before a fire made from the one lump of coal left in the coal-house.

The more we mortals build the more there is for the fates to destroy. I often wonder what urges us to acquire and to hoard. I do not believe we pile up possessions for the pleasure which they give us, nor even for the power, but for the hallmark which they set upon our success. And who is the man who dare define success? The very old, the storm-battered who have crept to harbour, are too full of bigotry and bitterness to have sound judgment, and the very young grasp at tinsel and the glittering trivialities, while those in the fine prime of lusty strength are in the thick of the struggle for the success which they would not recognise. They cannot see the wood for the trees. For success lies in the simplicity of the mind. The car-penter who makes a trim chest is as great a man as the millionaire who doubles and redoubles his fortune. When we understand rela-tivity, man's blind upward struggle, which thwarts itself and turns in upon itself, will at last be directed to steady constructive effort. Material gain is often balanced by spiritual bankruptcy. A man must beware that he does not exchange simplicity for satiety, that he does not loan his soul to commerce, which will pay him a bank balance by way of interest. No man can acquire a disproportionate share of prizes from the swings without losing on the roundabouts. Whom the gods wish to enrich they first destroy.

At any rate, as I crept shivering to bed that night I knew that my petty little scheme of expansion had been found out and frowned upon – seven feet deep.

Early next morning Esmé and I ate some bread and cheese in the freezing kitchen and dried out our cold, damp working clothes. We went with John Davies and Thomas to see a place where twenty-five sheep had been buried together. These sheep had managed to slip through the mountain wall, and were found at the foot of the wall which ran straight up to the mountain wall from the road, and which formed the west boundary of the ffridd. The snow cavern was plain to be seen. It was very small, and lined with ice where the heat of the bodies had thawed the snow temporarily, until the temperature had frozen the damp walls. The floor was fouled by droppings. The snow itself is quite porous, and sheep breathe easily many feet under; and here, too, they had been helped by the air which had filtered down through the chinks of the drystone wall.

And suddenly Esmé, who was poking away about a yard from the hole, vanished with a startled cry. I peered into the cavity into which she had fallen, and saw in the gloom the bright eyes of sheep. Speedily we pulled out five, and they careered off, full of vitality after four days' imprisonment, until they reached a bare patch a few yards away. There we had to leave them marooned, without food. There was nothing else we could do.

The whole situation seemed to call for violent decisive action, but till this discovery I had been at a loss, keyed up, but with nothing on which to vent my energy. The wall ran uphill at an angle of one in two. It was six or seven feet high, and here and there it was lost under the surface. The snow was loose, and as we moved we broke through to our waists, and often to our necks.

Thomas and Davies took one side of the wall, Esmé and I took the other. After three hours we reached the mountain wall, half a mile away, and we had found another twenty ewes. As we stood on the top of the mountain wall at a place where it stuck out above the snow, we were all silent, appalled by the same thought. Esmé first voiced it.

'There might be a thousand sheep buried along this wall!' The mountain wall is about six miles long. It is higher than I,

and I am over six feet, yet it appeared only here and there through the drifts.

'We shall need long poles to prod with,' I said, 'and it takes so long to get up here that we shall all have to bring food and stay up till dark.'

In the end we decided to go down again along the same wall where we had just searched. And almost at once I fell on top of a solitary ewe, buried ten feet from the wall. We began to feel like pygmies confronted with the task of gods. If the sheep were buried away from sheltering places, how on earth were we to cover the four square miles of Dyffryn probing with long poles?

At daybreak next day we started up the mountain wall where it began its climb from the road. We had with us nine-foot bamboos, and we prodded along the wall and quite a way wide of it. Every now and again someone would feel a yielding body, and we would dig out the victim. Often we had to flounder through the snow with the ewe on our backs for a quarter of a mile before there was anywhere to set her down. And lambing time was only three weeks away. I began to think that even if any of the flock were alive, the lambs would be aborted. We men used logic and thought the task impossible, for experience showed us that it was necessary to thrust the bamboos at nine-inch intervals on either side of the wall. But Esmé had no use for logic, and shamed us into following her.

And after a day or two the dogs gave up. Just at first they had shown interest when sheep had been dug out, and soon they had begun to hunt on their own, digging like mad things where they smelled a body, but the exertion of moving through the soft snow began to wear them down, and their discoveries soon lost the interest of novelty.

On the ninth day of the snow Davies told me that a big neighbouring farm had given up their hunt. There is a superstition in Wales that a buried sheep does not live more than nine days. I suspect this arises from the fact that no man wants to work at such intensity as ours for more than nine days. But Esmé had no use for superstitions, and by now we were her blind slaves. She drew on some mysterious reserves of nervous energy, constantly wet through, frozen with cold, returning to the tireless, comfortless house at dark only to snatch a meal and to stumble to bed, where she sank into a dead sleep, to awake by instinct with daylight.

As for us men, we began to get our second wind. The work was kill or cure, and none of us was broken, so that in time we became unconquerable. And after a while we began to have a little outside help.

By this time gangs of men had dug the road clear from Bettws-y-Coed nearly as far as Dyffryn, and one day from our perch up on the mountain we saw a car stop on the road below the ffridd. Two figures appeared, and began to flounder up towards where they could see us as black specks against the white. It was nearly two hours before they reached us, and both the newcomers were speechless. Our own bodies had become as hard as iron, and we could not at first understand their exhaustion.

But Leslie and Aubrey turned out to be good friends, and day by day, as they stuck at the work, they too began to grow more fit, until at last the nips at the whisky which they found it necessary to carry for medicinal purposes almost ceased.

But I was always relieved when Aubrey was safely down each night. He weighed fifteen stone, and had he really become stuck in the snow nothing would have got him out, and I expect he would be there yet.

The mountain wall became a nightmare to all of us. Tired as we were at night, we would dream of it. It was reptilian in its malevolence, curving in and out of the snow as a sea-serpent arches its way through water. And after ten days of starvation the sheep were becoming weak. It was no good with many of them just to leave them exhausted on a barren patch of bare, frozen ground. I thrust two broom-handles through a sack and made a stretcher, and we began to carry the weak ewes down the mountain to put them in warm buildings and feed them on egg and milk or gruel. Even when the ewes revived – and many died – there was nowhere we could put them to graze, and so we fixed up racks in the stackyard and kept them filled with hay. Many sheep, true to their mountain custom, would not come to eat, and we had to turn these out to scratch down through the snow for the dead herbage, or to snatch at sprigs of whin and heather, which thrust through the snow like dead fingers.

No sooner had we carried down one burden than high above us would echo a thin shout, and a tiny figure would gesticulate. Back we ploughed up the mountain for another load. We now began to find sheep dead under the snow. This was usually only when several were buried in one hole. The weakest would then be trampled underfoot and suffocated. Often we found sheep who had begun to pluck at the wool on their sides, chewing it for the grease in it, so that they appeared as if partly shorn, like French poodles.

For many days none of us dared to think what might be happening at the far end of the farm, three miles away. It seemed that what we could do was so inadequate that we might just as well carry on where we were. One day my curiosity got the better of me, and I started at dawn to plough along the buried road to the farthest boundary. I travelled at the rate of a mile an hour, helped by stretches of road swept bare by the wind, and at last was at the foot of the tremendous slope of tumbled rocks which swept up steeper and steeper above Cwmffynnon, until it reared up into the precipices of Glyder Fach, down which the German had dragged his broken leg. High up the slope, usually so broken, was smooth, buried under snow of tremendous depth; but lower down some freak of the contour had played the wind so that a big stretch was clear of snow. And sheep were here in scores and in hundreds, black against the sprinkling of snow which was dusted at their feet like powdered sugar.

Luck and I made our way up to them, and they stood or lay in apathetic groups, not troubling to move, or even to follow our movements with their eyes, as we circled in and out among them. But at any rate they were alive, and seemed strong enough to last a long while yet. Never before was there so drastic a demonstration of heredity. For countless sheep generations the flock had been bred for hardiness, otherwise not a tithe would have remained alive.

Next day the two amateur helpers and I set off back with sacks of hay. But no sooner did we spread it in rows on the ground than the prowling wind discovered it and whirled it wisp by wisp away. We began, the three of us, to prod our way from that end up the mountain wall. Already Esmé and her party were in sight a mile away. It was now a terrific task to get really weak sheep to succour, and we had to content ourselves with carrying bad cases to the big bare piece of ground, where we gave them sustenance from thermos flasks, and as we went back at evening we would carry with us the worst of the cases.

On the twelfth day the two parties met. We had dug out about two hundred sheep, and I think we all knew that we had Esmé alone to thank for shaming us into an attempt which had at first seemed impossible. As we sat we could see sheep, dark against the white, on every patch of clear ground, in twos, in dozens, in scores. It was obvious that the bulk of the flock was safe. The snow could not lie for ever. March was well advanced, the sun was growing in strength. The reserve vitality of the sheep was so tremendous that we knew they would stay alive almost indefinitely.

'Tomorrow,' said Esmé, 'we'll start on the cross-walls and gullies.'

But when we awoke on the thirteenth morning the windows were plastered heavily with huge flakes, and outside the snow fell relentlessly, blotting out every sign of earth as it piled visibly, flake on flake, undisturbed by a breath of wind. The countless sheep marooned the day before on their islands were soon belly-deep; then one by one the little groups were submerged, like survivors of a wreck struggling against the waters.

It snowed all day, and as it snowed we laboured in the gullies. Again but for Esmé we should have given up. But she insisted that we carry on the search for the sheep which had been buried under the first fall of nearly a fortnight before. When we found any it was difficult to know what to do with them. We could only rescue them from perhaps ten feet of snow to reward them with three feet, but we were able to give them gruel and to put them in resting places where we could find them again, and the worst could be taken to the buildings.

We had no sense now as at first of exhilaration in defying the caprice of the gods. We were weighed down by the knowledge of the inexhaustible strength against us. We struggled only because man's heredity is battle, but we struggled like blind creatures, fighting, dazed, by instinct rather than by reason.

But the gods must have been chivalrous, and as they gazed down on Esmé's puny, indomitable struggles and saw the inspiration which she had given to us her slaves, they felt suddenly ashamed of their malicious teasing, and on the fifteenth day the skies were blue, and under a warm sun the loose snow melted away into the hidden streams. Our blind gropings became purposeful again. We felt we had a chance to win, for man always attributes to his own efforts a relaxation of the fates. Hour by hour the brown winter colours spread, like a tea-stain creeping on a white tablecloth.

On the eighteenth day the ground below the mountain wall and along the bottom of the valley was clear except for the gullies, which were choked with snow, and except for the drifts, packed hard against rocks and walls. The sheep miraculously revived. It was as if they had rendered themselves dormant during their trials and were now waking. I do not believe that the hibernating instinct is dead in those animals which appear to have lost it.

We had now only the last drifts to search. The huge stretches over which we had crept were again firm earth. The spring grasses looked greener than ever after their protective covering. It seemed a fantasy that there had ever been trouble as we swung easily along. The sheep moved warily away from the dogs, eating as they went, as if they realised that they had many lost meals to recover.

On this eighteenth day I found my last buried sheep. As I trampled the wet snow which still lay heavily between the banks of a deep-cut stream I found eleven ewes. Six were dead – smothered, I think, by being trodden underfoot. Five were alive, and one so vigorous that when daylight opened over her head she took a standing jump out of her tomb and ran off.

And on the twenty-first day Davies found his last sheep. She was buried under an overhanging rock, and lay on her side in a pool of mud and water. He lifted her out and gave her gruel. Both her sides were bare of wool, which she had plucked in her hunger. After the gruel she got to her feet and tottered away. A fortnight later she bore a lamb, and subsequently reared him. The increase which I had gained over my numbers at the valuation was gone, and my denial of many things to keep sheep rather than sell them had been vain. But humility was a lesson learned. And Esmé's courage became an example, for the farm which had given up on the ninth day had lost a third of its flock. One day at the very end of March Esmé and I went for a walk, determined to enjoy the snow which had been our tribulation. We climbed to our boundary ridge, which was still deep in hard-packed snow. The sun shone brilliantly. Tryfan showed black against the white-draped background of the Carnedds, her sheer rock flanks too steep for snow to lie upon. The range of Snowdon glistened against the deep blue sky, and its height was accentuated by its brown-green foothills. We climbed Glyder Fach, swung along the ridge to Glyder Fawr, and began to descend to a deep hollow where lies Llyn-y-Cwn, the 'Lake of the Dogs'. I squatted on my heels in a glissade. The quickening slide became wildly exhilarating. The snow changed to ice, and I spun wildly on my back as I slipped ever faster towards the fierce black cliffs which overlook the lakes. Far away Anglesey swung in an orbit as I plunged, and away to the side of Lliwedd the blue of Cardigan Bay melted into the sky. On the edge, right on the edge, I stopped. Man's last instinct is self-preservation, and he has to love indeed to save another at his own expense. But my subconscious mind had prompted me, and I found myself, bruised, breathless, and quite apathetic, clutching a boulder at the crest of the cliffs. And that boulder rocked loose in its socket of ice. It took me twenty minutes to creep back up the ice slope to where Esmé, pale faced, waited. It needed great force to kick a niche in the frozen crust, and it seemed to me that any violent movement must shake me from my delicate hold and set me on a relentless, ever swifter journey to the lake five hundred feet below.

That day I was not shaken badly, but Esmé had great difficulty in following me. We tried to come down through the crags that overbear Llanberis Pass, and every route seemed barred by ice. It was too late to return the way we had come, and it is doubtful whether we should have survived a night. But as the lights began to twinkle on in the valley we slid off the last patch of snow. There were six miles to go, but grass was underfoot.

Next day came the reaction. We climbed Snowdon up the Watkin track, and when we came to the smooth dome of the peak we found that it wore a cap of solid ice. It was a place for axes and a rope, but I found myself part way up, perched like a sparrow on the housetops, afraid to return, unable to move on. We had part circled the peak as we climbed upward, and between my shaking legs I could see the shining ice bend ever sharper downward, until it vanished into a void where the far shores of Llydau showed a thousand feet beneath. Had we slipped, no earthly power could have saved us. My fingers felt at the smooth, hostile slope, and my boot-nails barely scratched the surface. I stayed where I was by so little friction that I verily believe heavier clothes would have sent me plunging down.

And as I broke into a violent sweat in the bitter wind, Esmé crept past me on all fours. She was kicking her toes into the crust so that they had perhaps a half-inch of purchase, and the palms of her hands rested on the ice. I think that when I followed her I won the greatest moral victory of my life. In such a position one is so alone, so beyond the hope of aid. I fell into the summit hut, which was half filled with snow blown through its broken door, and shook like a palsied man.

And the only other people we saw that day were four Alpine climbers, roped, with ice-axes, who had had difficulty on the ordinary path from Pen-y-Pass. They thought us mad, but we were only ignorant, and I believe the gods condone ignorance where they deplore calculation.

I think reaction from my experience of the previous day had overtaken me. Such reaction is quite often delayed. I dimly remember a wild scramble to get below the snow-line before dark. Even Esmé ruled out the way we had come, and the ordinary route by which the climbers had appeared. It would have been possible to descend by the track of the funicular railway, but that would have brought us down many miles from home, and my manhood must subconsciously have resented this lubber's hole. We tried one ridge which led off towards Beddgelert, but it sharpened into a knife-edge of ice, and I could not bring myself to straddle it. We regained Snowdon summit with great difficulty. The peaks around us were not hostile – just indifferent. They had at a whim called on the snow to sport with our flocks. Now we were labouring the attention they had given us, and they were bored. We could live or perish as we wished. For their part they took no further interest.

We lived. We found a way down along the crest of the Clogwyn ridge, and walked home nine miles in the dark. An Alpine climber would have shuddered at our foolishness that day, and at times even Esmé's hard little body had trembled a little.

As we trudged home through the narrow defile of Llanberis Pass, the white summits shone in the starlight, as cold and lifeless as the mountains of the moon.

7

The Fox Shoot

AMONG THE ROCKS AROUND DYFFRYN VALLEY the foxes make their dens. They are there in hundreds. I am sure that two or three hundred are killed each year on the farms around Capel Curig. One neighbour of mine, a white-haired old farmer who spends much of his time hunting the vermin, once killed over a hundred, including cubs, in a period of twelve months. It is difficult to tell how many lambs the foxes destroy each spring. I expect it is safe to say that they take fifty a year from Dyffryn alone. There is little game on the bleak hills around Capel Curig, and the foxes themselves have discouraged poultry-keeping, so when the vixen needs food to support herself and her cubs in the spring, the newborn lambs are a godsend to her.

It may not be only lambs which the foxes kill. An old man once told me that one winter day he was following the tracks of two foxes which showed fresh in the snow. He was hoping to trace them to an earth, where he would bolt the foxes with his terriers and shoot them as they ran. And suddenly he came on a full-grown sheep dying in the snow. Fox tracks were around her, and there were marks of a struggle. Her throat was so badly torn that he finished her off with his knife. He must have disturbed the foxes, for he never came up with them that day.

There is nothing which so angers a farmer as to find decapitated lambs wantonly killed for sport after they have been safely reared through the many natural hazards of their mountain birth. The farmers around Dyffryn have tried many ways of keeping down the number of the foxes. At one time strychnine was often used. A slit would be made in the stomach of a lamb just dead, and the poison when inserted would spread through the body, rendering fatal a bite on any portion. But strychnine is a tool which has a way of recoiling on its master, and many a good dog has been killed instead of the fox, for there are few farm dogs which will not pull at carrion. Thomas Davies has a dog which never eats at home. There are always carcasses to be found lying about in the hills, and Moss prefers to find his own food. Very often, too, a neighbour's dog discovers the strychnine, and, as dogs are difficult to replace, bad blood will be bred between adjoining farms. There is no room for quarrels between neighbours, because every farm needs help at the gatherings. It is possible to warn everyone to keep the dogs in while the bait is left out at nights, but this method is clumsy, and in any case the fox seems to know what is unsafe for him to eat. I believe it is true that strychnine kills seven times. The fox eats the lamb and dies. The crow pecks at the dead fox and dies. The dog sports with the dead crow and dies. There is no end to the menace. Just recently the sale of strychnine has been restricted. It is a good thing.

The same objections apply to traps. A favourite way to trap a fox is to make an island in a little pool, with just one dry pathway to the island. The bait is left lying on the island, and a gin is concealed on the path. A fox hates to wet his feet, and will have to pass over the trap to reach the bait. But a dog may get there first.

Then down in the Glaslyn valley, ten miles from Dyffryn, is a pack of foot-hounds. Now and then they hunt our country, but they are more picturesque than useful. In Cumberland the hill packs kill many foxes, but their country is more grassy; and there are fewer foxes to distract the scent. Dogs cannot travel fast over the more rugged Welsh country, and I am sure that they do not chase the same fox for more than a few minutes at a time. The fox takes a little exercise, then changes with another, and the weary dogs are left to toil after a fresh quarry.

The only successful way to kill our foxes is to surround their dens with guns, and to bolt the vermin out with terriers in the way ferrets are used with rabbits. As fast as we kill the foxes they are replaced by new recruits who move up from their secure nurseries among the plantations of the Forestry Commission around Bettws-y-Coed. The plantations are so impenetrable that it is doubtful whether dogs could force their way in. Men certainly could not. So when the Forestry people give us their permission to hunt the plantations we have little to thank them for. However, we still shoot the foxes from a mixture of sport and necessity, even though we know that the gaps in their ranks will speedily be filled again.

During one lambing season at Dyffryn the loss of lambs through foxes was particularly bad. Everyone else was complaining too, and down towards Llanberis many farmers were driving their flocks close to the house each night and putting lanterns out in the fields. But that did little good. And on the bigger farms it was impossible to crowd all the sheep into one place without starving them, for small enclosures were soon cropped bare, and the time available to the ewe to graze and to build up her milk supply was drastically cut.

At last a series of intensive shoots was arranged. Esmé and I were ceremoniously invited to one. We were as honoured as if it had been an invitation to a royal garden party, for we knew that we were accepted by our neighbours. Early one morning we met the party near the twin lakes. It was a Saturday, and some of the quarry-men from the little slate quarry at Capel Curig had managed to take the day off. Three quarrymen turned up, six farmers and a keeper. They had with them eight Welsh terriers and two sheepdogs. The leader of the party, who was, I think, self-elected, was Harry Parry Parc. He was a little, comical man, like a cobbler in a fairytale. He had been christened Henry Parry, but there are many Parrys in Wales, and as he had once worked at Parc bakery he was known as Harry Parry Parc. All the men were armed with guns, mostly rusty hammer weapons of uncertain reliability.

It was decided to try first a series of known dens high up on Siabod. None of the farmers objected to the clearing of another man's land of foxes. This generosity sprang from nature lore, not virtue, because they well knew that a fox tends to kill away from his home. A Siabod fox would be most dangerous on neighbouring hills.

We were close under Siabod, and its long crest was hidden behind the bulge of the lower slopes. The ragged edge of a cliff cut the skyline, and among the debris at its foot were the dens. We began to climb towards the cliff on a long, upward slant. Esmé and I walked behind the party with one of the quarrymen, who told us many stories of a fox's cunning. This quarryman lived for fox-hunting. He loved the wild parts into which its pursuit led him, the rare plants and ferns he saw in unknown places, and the queer doings of the creatures he came across among the hills – buzzards, curlews, ravens, choughs, hares, badgers, and sometimes the almost extinct pine marten. On his quarryman's wage he paid for the licences and the food of a pack of five Welsh terriers. They were aloof, hard-bitten little brutes, scarred with the bites of many an underground battle with foxes. But he and they were devoted friends.

Some time afterwards this quarryman was drilling a rock for blasting. When the hole was partly made he changed the bit of his drill. The new bit was a shade smaller in diameter, and so when the hole was finished a little ridge was left halfway down. He poured in the powder, and began to tamp it down with a brass rod. But the rod struck the little ridge, and somehow there was an explosion. It was the first accident at the quarry for many years, and even today no one is quite certain how it happened.

The quarryman still goes out fox shooting. The skin of his face is blue with the slate-dust driven into it, and his uncertain eyes peer dimly through spectacles. He fiercely resents a helping hand, but he rarely stumbles. The mountains seem to look after him. I wonder if those cold goddesses melt before helplessness where they would break strength.

Harry Parry Parc led us upward for half an hour before he called a rest. We all sat down except the keeper, who was taking a busman's holiday from a nearby estate. He went off by himself, and sat on a rock with his pipe. The rest of us chatted away, until the terriers suddenly started a fight. The quarryman's five terriers attacked the other three, the two sheepdogs dived in on top of the terriers, and Harry Parry Parc dived on top of the sheepdogs. Of course, an anonymous dog bit him on the ankle, and his yell was so loud that the dogs were startled into quiet.

Everybody enjoyed the diversion except the victim and the keeper. The keeper had not even looked round. But Harry Parry Parc was not easily to be deterred from leadership, and presently he shepherded us to our feet and limped in front of us towards the cliff.

The cliff formed a small amphitheatre. The arena was choked with the debris which the storms of years had loosened from the parent cliff. Most of the stones and slabs were huge, and it seemed an impossibility to search among the hundreds of holes and caverns. But the terriers knew what they were about, and overran the place in a moment, all working quite separately from one another. When a dog had quartered an area and found nothing of interest, he left it and busied himself with another plot. In the centre of the arena a tilted slab lay over two huge boulders, and in the cave underneath many cracks and crevices led away underground. Again and again the little dogs sniffed at the entrance. They seemed to weigh the evidence a moment before seeking elsewhere, as if they were undecided about the cave and hoped for more certain traces in some other place. But one by one they came back to the same spot, and suddenly one led the way in.

At once Harry Parry Parc split the men into two parties and sent them around the wings of the cliff, taking the two sheepdogs with them, until they could line the crest and command the whole arena with their guns. Noise did not seem to matter, for they were all chattering to one another in the most animated way, except for the keeper, and in a little while every one forgot about him because he disappeared. I wanted a closer view of the hunt, and stayed down below with Esmé. We scrambled on to a great boulder below the cave and watched Harry Parry Parc arranging his men along the cliff. In the centre of the crest of the cliff stood a little pinnacle of rock. Harry Parry Parc mounted the pinnacle, and stood, gun in hand, striking the attitude of a last descendant of a long line of generals about to die leading a forlorn hope. Every now and again he scrutinised his wounded leg.

From my boulder I could look straight across the valley to the slopes of Dyffryn. My new-found love of the place was still fresh, and pride brought a lump to my throat as I watched cloud shadows chasing one another across the sleeping body of Glyder Fach. Just peeping over the shoulder of the Glyder was Tryfan. The river flashed in a thin streak along the sunlit floor of the valley. A crawling car was balanced on the tightrope road. In those days I did not understand the siren loveliness of the mountains. I thought them kind and beautiful. Now I wonder sometimes if they are only beautiful, and yet I love them still. Esmé poised herself beside me like a windswept elf, and truly as we smiled at each other in understanding we would have changed places with no king or queen. Then from deep under our feet came a muffled barking. No one knows how far the cracks run under the rocks. Terriers often jump down into places from which they cannot return. Day by day their cries grow fainter, until their faithful hearts fade with the last whisper. The owners try their best to reach the prisoners, but crowbars, spades or even dynamite will not always free the dogs. I have known relays of men dig almost day and night for three days to rescue a dog. And so they should.

The barking underneath the rocks was very shrill and excited, but so faint that Harry Parry and his gang did not hear it. I slipped off my boulder and tiptoed to a fissure from which the noise had seemed to come. I put my ear to the crack, and thought I heard sounds somewhere within. But outside was a hush that I could feel. The tenseness in the air startled me into looking up, and I found myself staring at a semicircle of levelled gun-barrels. If some one had cried 'Fox!' at that moment my head would certainly have been blown off. Harry Parry Parc was quivering all over. I undulated away from the fissure like a snake, and climbed back sweating on to our rock.

And suddenly it happened in a flash. A red shape appeared as if from nowhere, and slid gracefully downhill through the rocks. The fox was away in an instant, yet his movements seemed quite unhurried. I grabbed Esmé, and we threw ourselves flat on our faces at once as a fusillade roared from the top of the cliff. Pellets ricocheted and whined all round us. When the fox had broken cover Harry Parry Parc had been standing on his pinnacle on one leg to ease his bitten ankle. As he swung his rusty hammer-gun it went off, both barrels at once, while still halfway to his shoulder. The butt took him in the stomach, and he and his weapon vanished abruptly.

The terriers popped up one by one from various crevices and stood a moment panting and blinking in the strong light, before they propelled their underhung little bodies along the scent. Stumbling and yipping with frantic excitement, the two sheepdogs tore down the cliff as they followed their quarry by sight. But the fox was two hundred yards ahead of them, and seemed to be ambling as he turned uphill around the flank of the cliff. Quite soon the dogs came back. They had made their gesture, and it was obvious from the way they looked at the men that they considered the shooting purely a gesture also.

The men had scrambled down into the arena by this time – all except the keeper, who had long since been lost sight of. They were very pleased with themselves. Each one reckoned that he had put two charges into the fox – even Will Pen-y-Bryn, until some one pointed out that his gun had a single barrel. Harry Parry Parc arrived last, with a cut chin added to his limp. He said that he had fired as quick as lightning, and had distinctly seen fur fly from the fox's ribs. To prove his point he went off to look for the fur.

I remarked that the fox would be crushed by the weight of the lead before he could run far, and suggested going to search for the body. But nobody considered this necessary. They said that the fox was dead right enough, and that it was of no use to waste time beating about in the heather.

And then from somewhere behind the cliff where the fox had been last seen there was wafted a single dull report. We gazed on one another with a wild surmise. In a little while a solitary figure came plodding along the top of the cliff. It was the taciturn keeper. A shapeless object was slung over his shoulder. When he was opposite to us he flung his burden over the cliff, and a dog-fox landed with a thud in our midst. The keeper sat down and lit his pipe.

Harry Parry Parc poked the fox about with his foot. 'I knew I hit him!' he said.

Later that day we shot a vixen as she came out of her earth. Her teats were full of milk, and one could squeeze out the milk from the dead body. The men dug away at the earth, but could not reach the cubs. However, it was not necessary. The cubs would never trouble us. This last kill made me a little sick at the time. In those days I had not learned how dangerous and how cruel the fox is. Now I should have no compunction in wringing the neck of the most appealing cub. There is no virtue in the fox, only evil.

Perhaps it is not true to say that the fox has no virtue. If a vixen is killed when her cubs are young, the dog will do his utmost to provide for the little ones. A fox cub can do without milk at a very tender age, maybe from the fourth or fifth day, if he has to. So the widower is sometimes able to rear the little ones by himself.

A fox cannot carry lead. Like the gangster he is, he dies of lead poisoning. He is such a foul feeder that his wounds turn septic, and it is said that if a single pellet from a charge of shot penetrates his skin he will die. There is a market for live fox cubs. They are bought and reared as pets. They become quite tame, and will grow up to be friendly with dogs. I once walked into the bar of the Falcon Hotel at Stratford-on-Avon and found a half-grown vixen regarding me from the depths of an armchair. Many wild animals lend themselves to domestication. Once in Canada when I was up in the woods of Northern Quebec I came upon a settlement where the innkeeper had a tame black bear running about the place. He had found the bear when it was but partly grown caught in a trap. The animal had not been much injured. There was a veranda in front of the rough wooden shack, and stuck in the ground about twenty yards from one end of the veranda stood a wooden post. When the bear was first captured the innkeeper had kept it tethered to this post.

One day a party of Americans turned up after an unsuccessful hunting expedition in the woods. One of them, loath to return home without a trophy, offered fifty dollars for a shot off the veranda at the bear. He missed, and the bear is still at La Barrière for all

I know. It is curious how an inexperienced hunter misses game, though he may be an excellent shot. I have seen hard, tough men behave like schoolgirls at the sight of their first deer as his white scut flits through the brush. If he stops, as he sometimes does out of curiosity, the hunter will miss a standing broadside shot. This phenomenon is well known up in the woods. It is called 'buck-fever'.

There are otters living along the river which flows through Dyffryn valley, but it is only too rarely that one sees their sleek bodies as they sport in the water, and when one does see them it is often only as a shadow in the river at dusk. Not very long ago Thomas saw a young otter padding across the bottom meadows. He threw his coat over the beast and caught him. The otter was very beautiful. He had a broad, flat head rather like a badger's, and very bright, intelligent eyes, which looked up startled but unafraid. He had a rich coat like sealskin, and his flat pads were as expressive as human hands.

Esmé fell in love with him at once, and we decided to try to tame him. There is a pigsty behind the old farmhouse, and we put him there to sleep for his first night. The walls of this sty are five feet high, and are well built of stone. The gate is made of steel, narrowly barred. We laid sheets of corrugated iron flat along the top of the walls so that they overhung the sty, to prevent him from scrambling up the wall, though that seemed impossible. In the morning he was gone. Hedgehogs and tortoises have a way of vanishing from escape-proof enclosures, and most animals seem at times to have supernatural powers at their call. But to this day we do not know how the otter left his sty.

Secretly Esmé and I were rather glad that he had gone. The otter is the very embodiment of wild, careless freedom, and if sometimes we catch a glimpse of him or of his brothers when they do not suspect our presence, we are more than recompensed.

8

The Cutting

BY THE MIDDLE OF MAY the lamb crop is successfully reared, in spite of the foxes and the snow. The time comes to turn the whole flock back to the mountain, but before the lambs go up with their mothers we castrate the ram lambs and earmark both ram lambs and ewe lambs. We manage to complete the work in one day, and as there is so much to do in the pens, just as many men turn up as for a mountain gathering.

We deal with the flock in two portions, which we do not mix. A lamb is very quick to find his mother again after the couples have been shuffled about in a pen, but in these early days it is safer to limit the size of the flock among which he has to seek her. So we treat the two-year-olds from below the road as one unit, and the three- and four-year-olds as another. The two-year-olds and their lambs are first gathered together out of the meadows and are driven into the pens. Young lambs have no cohesion. They do not follow the popular movement, and when once they have lost touch with their mothers in the press they bolt back past the dogs, twisting and leaping like rabbits, until the men have to catch the last few by hand. And that is a very arduous job.

In a large flock there are always about equal numbers of lambs of either sex. We castrate the ram lambs because, for one thing, we need only one ram to fifty ewes. Thus nearly all the ram lambs will be sold off at the annual September auction sale on Dyffryn, and will be bought by lowland farmers, who will fatten them for the butcher. Most of them are eaten before Christmas. Therefore, as the rams are going to be killed for Welsh lamb, we castrate them and turn them into wethers. As wethers they grow more fleshy. and do not dissipate their energies in following instincts which are now prohibited. For animals are very precocious, and when the lowland man buys the wether lambs and takes them home with him they are more tractable, and stay readily enough in his fields.

In the old days before the Great War the wethers used to be kept on the mountains until they were four years old, when they were sold off just as draft ewes are today, their places being filled by the spring's crop of wether lambs. But the big joints of wether mutton are out of fashion now because of small families and a diminished national appetite for meat, so that it pays best to keep a ewe instead of a wether, and to breed from her. The old-fashioned hill farmers still keep a few old wethers. They reckon that the hardy old troopers keep their highest, bleakest boundary for them, and discourage the trespass of strange sheep. But at Dyffryn we have bred our ewes to be just as hardy as the wethers ever were, and their whole instinct is to press upward to the limit of the land. The one real loss from the lapse of the old custom is the heavy crop of wool which the wethers produced. Farmers used to pay their rent with the receipts from the sale of wool, for the wether's fleece weighs as much as six pounds, against the ewe's two and a half.

The old method of castration was crude. One man would sit on a bench, the lamb held between his knees, while his partner, using a sharp penknife, slit the bag and drew out the two testicles one at a time, severing the cords that held them. We used this method when I was first at Dyffryn, and I used to be rather appalled at the big heap which slowly piled up, a gruesome monument to dead virility. It is a sickening thought that in ancient times, in the days of the Assyrian, the Persian and the Roman Empires, this same crude scene must often have been enacted, but with proud captive warriors in place of lambs. Lamb's fry are said to be a great delicacy, and I know of a hotel which will pay handsomely for them.

But an Italian has now patented an instrument known as a bloodless castrator. The tool is rather like the broad-headed pincers which blacksmiths use to extract nails. A salesman first demonstrated to me by folding a piece of paper over a length of string. He closed the instrument on the paper. The string inside was severed, but the paper was uncut – only creased a little. The bloodless castrator works in the same way on the lamb, or the grown ram or the bull for that matter. The cord which holds the testicle is cut, but the bag remains unwounded.

I bought the implement as soon as I had seen it work, and have used it ever since with complete success. The operation is not nearly so drastic for the lamb as was the old way, and the testicles shrivel away and do not seem to give pain or to retard growth. The old rough way was particularly dangerous on hill farms, because a late frost coming soon after castration would kill many lambs.

One particularly barbarous means of castration is still sometimes used in outlandish places on old rams who have had their day. A short length of stick is bound tightly across the top of the bag, which is simply left to atrophy. The process takes months. We work fast at the pens, for there is much to do. On either side of the gate from the sorting pen a man is seated on a bench. One man holds the ram lambs for castration, the other the ewe lambs for earmarking. Three or four of the most active men are in the pen itself to catch the lively lambs. Thomas, Gwylim Pantffynnon, and Will Hafod all have the competitive spirit highly developed. They will spur one another to tremendous feats as they gather in their arms three or even four of the sturdy lambs at one time. Often they flood us people outside the gate and yell derisive abuse at us. Old Bob Henblas always comes himself on this day, instead of his shepherd. It is traditionally his job to earmark the ewe lambs, and his bloodstained, gnarled fingers snip and slit the ears with machine-like precision. Meanwhile three of us are at the castrating. Little Davydd Pentref usually sits to hold the ram lambs, and he regales us with endless anecdotes about his surprising misfortunes, which are all of a comical turn. Davydd holds the lamb upright, its rump on his knees. A second man stretches down the testicle so that the cord is taut for me to cut with the bloodless castrator. We change sides and operate on the second testicle. The whole business takes only a few seconds. Then Davydd puts a notch in the lamb's left ear, before he lets him gently to the ground. The full earmark is not necessary on the wether lambs. The notch is quite sufficient for my neighbours to recognise, though the smallness of this identification mark used to worry me greatly at first. I used not to realise that the least slit in a sheep's ear is as conspicuous and as individual to a Welshman as, say, car radiators of different marques are to a garage hand. The main reason for avoiding the full earmark is that these wether lambs will be sold within four months or so, and the new owner may wish to impose his own mark on them.

We do not find many lambs which are fit to keep for rams among the first batch in the pens. For these are the two-year-olds and their firstborn. A young sheep never produces such sturdy progeny as her more mature sister in the ffridd.

At last the lambs are done with, and we sort through the ewes. Here and there we find one who has still to lamb, and we pick her out to go back to the meadows. Meanwhile John Davies will be busy clipping wool from the tails of any ewes who have been scouring. This is a precaution against maggots. In warm, sultry weather the bluebottles become active, and they like to lay their eggs in just such a moist, dirty place. The eggs hatch, and the young fly appears as a fat, white maggot about the size of a grain of rice. He lives on the flesh of his hostess, and will, if he is left to himself, eat her alive with the help of his swarming brethren. When the work is done we drive the whole flock away to the enclosure next to the ffridd where the ewes will be left undisturbed for two or three days to pick up their lambs, and the lambs to recover from their operations. Then they are all turned up to the mountain together.

As soon as the two-year-olds are finished we go up to lunch in the old farmhouse. Esmé takes great pride in her gathering meals at Dyffryn, and we find her preoccupied little person laboriously carving meat with a bread-saw off a round of beef as big as herself. When the day is cold and a fire flickers in the open grate the old dark room stirs in its slumbers, and the shadows creep from the corners to eavesdrop. The spirits of the place come to sit on the shoulders of their descendants, and Esmé and I are carried back many generations.

After lunch is finished the men jump to their feet with a clatter, as if at an unseen signal. I had noticed the same unanimity in the kitchen of Dyffryn house on my very first visit. We go out and climb the ffridd boundary to the mountain wall, before we move forward in a line, sweeping the sheep before us. There are a dozen men, instead of the four or five needed to gather the ffridd and we travel quickly. Now and then someone finds a late lamb newly born, and leaves him behind with his mother. In two hours the three- and four-year-olds are wedged in confusion against the entrance to the pens. There is a much bigger flock here than the batch of the morning. The frantic dogs herd racing lambs until both lambs and dogs are ready to drop with fatigue. It is half an hour before the sheep are safely shut in. Now the routine of the morning begins again. The earmarking and the castration go on endlessly.

But there is more interest now. We can expect to find lambs fit to keep for rams. Davies shouts, 'What about this flamer?'

All work stops. Thomas will light an eternal Woodbine. Gwylim Pantffynnon will drop two lambs and wade closer through the milling sheep. From the oldest to the youngest, everyone must express an opinion. The tail is too short. The coat is too long. The head is like a goat's. The nose is not black. The legs are like a camel's. The bone is too light. In the early days I used to look surreptitiously at old Bob Henblas as he came to lean bloody-fingered over the gate. He would nod or shake his head almost imperceptibly. I would at once astonish everyone by giving an instant firm decision. At once there would be a clamour of acclamation or despair. Some would say that the lamb was a paragon; others, horrified, that I was letting loose a blight among my flock. Old Bob would give the shadow of a wink and retire to his ear-slitting. Gradually I became competent to make decisions myself. At last I even picked one day a ram lamb against Bob's advice. It was a proud moment. I had my reasons; I could explain them. Right or wrong, I knew what I was doing. That we had to cut the lamb later on in the summer did not matter. Old Bob was very nice about it.

'I do only give my opinion,' he said at the time. 'And I can be wrong, like any other fellow. You keep the lamb. He may come better than none of them.'

Old Bob Henblas is no wet nurse. He likes a man to stand on his own feet, and I think he was glad to see that I was becoming confident enough to make technical decisions; for not so long before I should have been puzzled to tell a ewe lamb from a ram lamb.

In all we save each year about twenty ram lambs. As we see them from time to time at gatherings throughout the summer first one, then another is castrated as undesirable points appear during his growth, until in the end perhaps eight or ten ram lambs are left with their manhood. The older sheep are presently finished, and back they go into the ffridd for their rest before they take up summer quarters on the mountain.

It is pleasant to walk the mountain when the full flock, freshly lambed, is up. Esmé and I feel then that we do possess a birthright. There are maybe nearly three thousand head of sheep on the slopes. About one sheep to each acre. But Esmé soon caught my own wariness, and never talked expansively when the hills could hear. One can never know when some thoughtless phrase will penetrate their indifference and call forth a contemptuous reprimand.

But we do not expect many losses now. A few of the older lambs may die of woolly kidney, and a few more from misadventure, but the bulk of them should be quite safe. Woolly kidney is a difficult ailment to treat. There are no symptoms except death, and when that is diagnosed even the proprietary-medicine people agree that it is too late to do anything. I believe that veterinary circles are still a little puzzled by woolly kidney. They say that the disease is hereditary, and that the lamb is born with his kidneys in a pulpy state. For a little while this seems to stimulate him, for when he is found lying dead he is always one of the best lambs. Farmers often cut open one of these inexplicably dead lambs in order to investigate. In the stomach they find a little hard ball of wool. 'Ah!' they say. 'This lamb has died of woolly kidney.'

Probably he has, but the wool ball has nothing to do with it. Nearly every lamb will have wool in his stomach, swallowed by mistake during his vigorous pulls at his mother's udder. Old-fashioned farmers take this misapprehension so seriously that they go to immense pains to clip all around each ewe's udder. But the term 'woolly' comes from the spongy appearance of the dead lamb's kidney.

There is a very deadly lamb disease which is becoming dangerously prevalent, and whose origin is also rather debatable. This is known as lamb dysentery. It is preventable by injections of a serum, and, curiously enough, this serum mitigates too the losses from woolly kidney. Lamb dysentery is a fearful disease. Soon after birth a lamb begins to scour. He dies maybe in two or three days' time. The deaths can be prevented if every lamb is injected with serum before he is twenty-four hours old, and preferably before twelve hours. Each injection costs a shilling, and at Dyffryn it would be extremely difficult to find each newborn lamb. I dread the day when the disease will creep up to Dyffryn and will touch the hiding places of our rough country with its skeleton fingers.

The trouble with lamb dysentery is that the bacteria get into the ground. I believe that the ewe becomes a carrier, and that she infects the lamb during the foetal period, or through her milk after his birth. He soils the ground, and another ewe picks up the infection, which she will transmit to her own lamb that season or next. But when the time comes to send the flock back to the mountain we are safe from lamb dysentery for one more year, because a lamb can be considered immune after he is three weeks old. Farmers often behave like ostriches. I know of a smallholder who lost fifty lambs out of a hundred because he was afraid to call in a veterinary opinion to confirm his own suspicions. And farmers, too, are often ignorant. The symptoms of lamb dysentery are obvious and unmistakable, and the disease is a byword. But there is a mountain farm all too close to Dyffryn which suffered losses throughout a whole lambing season from this cause and considered them unexplainable. The disease has now got a firm hold on their ground.

So when Esmé and I survey our swarming flocks we do so surreptitiously, for the mountain gods are jealous. It is very curious how primitive one's instincts become in a wild place. The ancient Greeks used to call the Black Sea, notorious for violent storms, the Euxine, or Hospitable. This euphemism was a moral bribe to the gods. And we are just the same at Dyffryn. Indeed, I am not at all sure how much the annual harvest festival of the Christian

Church is not a pagan rite. One often renders devout thanks for a ruined harvest, presumably lest greater evils befall. At any rate, I am acutely uncomfortable when people throw up their hands in wonder when they see our summer flock gathered in one lot. Maybe I fear that I shall be taught my place by some further visitation like the snow.

9

The Pig Idea

ONE SPRING, WHEN SHE HAD been two years at Dyffryn, Esmé remarked to me, 'The hotels in the village must have a lot of kitchen waste.'

There are five large hotels that cater for the visitors who come to stare uncomprehendingly at the hills. I agreed that there must be a good deal of waste, and assumed that Esmé was in labour with one of her Ideas. She collects pamphlets on the queerest subjects, and I thought she purposed teaching the hoteliers how to make a Welsh goulash from old bones and potato peelings. But a day or two later she said, 'If we cut off the tops of old oil-drums we could make bins to collect the hotel waste.'

I thought then that she had come across a booklet dealing with the manufacture of a fertiliser compost from household scraps, but she set off hotfoot for our rubbish pit, and took with her a hammer, a hacksaw and a good chisel. Her efforts reduced several drums and the chisel to crumpled ruins. This inefficiency was subtle, for John Davies and Thomas, impatient as all men are at a woman's unhandy use of tools, took her weapons away from her and tidily opened several drums. Esmé fitted wire handles to them and returned proudly to the house. She then left a drum at each hotel, which she changed daily for an empty one. The stinking refuse was tipped away rather too near the house. Her pastime seemed unhealthy, but not dangerous. Presently she said, 'If I bought some pigs I could feed them on the swill.'

At last I understood what she was about. I asked why she had not bought the pigs before accumulating her typhoid trap in the yard, and she pointed out that she had wished to weigh the swill in order to find out how many pigs could be fed. She said that she proposed to buy twenty-nine weaners at eight weeks old, and to sell them after four months at a profit of twenty-one shillings a head. My remarks about guinea pigs were not received with any amusement. Esmé had written to most of the large firms who sold proprietary foodstuffs, and she proved her point to me by quoting a mass of figures from the pamphlets which they had sent to her. Most big foodstuff concerns keep large advisory staffs and run laboratories to aid their customers; and, since Dyffryn is a large farm and is well known, a spate of salesmen and experts began to arrive. I disclaimed all knowledge of pigs, and Esmé secreted herself with the visitors in the sitting-room to pick their brains. After that the disillusioned men would wander back to their cars, which were often hopelessly stuck on the steep hill leading up to the house.

The flush of salesmen subsided, and Esmé asked a friend of hers, a very lovely girl, to stay with us. One afternoon they took the car, a touring baby Austin, and drove to a local mart. I heard odd details of the expedition from farmer friends. Esmé's guest was dressed for the beach, and would have been attractive in the south of France. At the sale she caused a stoppage of business. The two girls wandered up and down the pig-pens and poked weaners to their feet with sticks, the better to view them. They compared their shapes with diagrams in a booklet on Bacon Types. The pigs protested vociferously at the disturbance, and the noise made selling so difficult that the auctioneer sent envoys to request a truce. This was granted, as Esmé now knew what she wanted. When the time came, the girls stood among the drab, black-avised hillmen like orchids in a cabbage-patch, and Esmé bid until she had bought twenty-nine weaners of Bacon Type.

A pig is very difficult to lift. Even a weaner weighs from thirty to forty pounds, and is both strong and slippery, and as soon as it is touched it makes such a noise that an amateur always lets go, self-conscious because of the attention which the piercing squeals attract. The auctioneer again had to send ambassadors to still the tumult and to assist in the loading. They managed to pack ten pigs into the Austin by removing the passenger's front seat. The pigs thus had free access to the controls, and Esmé's guest sat on the back of the hood armed with a spanner to discourage attempts at suicide by pigs who wished to hurl themselves overboard. A mob of sightseers, headed by a sergeant of police and an RSPCA inspector, escorted the shrill caravan through the gaping crowds. The two officials could not find anything really wrong, but wished to make an arrest on principle because of the unorthodoxy.

Esmé erupted into the sale yard twice more, and reduced the place to chaos while the car was loaded up again. And when I returned off the mountain at dusk the twenty-nine newcomers were eating a noisome brew by the back door of the house. John Davies is always impressed by Esmé's Ideas. He was watching with a pleased smile. He said, 'Indeed she have bought well, she have. See this flamer's thin ear? I do always like a pig with a thin ear.'

The pigs soon found their way to the vegetable garden. I began to learn how difficult it is to fence against pigs. If once a pig thrusts his nose under a fence, the body follows as a matter of course. In time I made the garden pig-proof, and forcibly discouraged skirmishers, who would stroll nonchalantly towards the fence when they thought the coast was clear in order to conduct sapping operations. When they discovered that it was a costly business to force the garden, the whole herd began to trot down the hillside to root among the hay-meadows. Even Davies baulked at this.

'The flamers have raise' tumps,' he grumbled. 'Them tumps will catch the mower-blades.'

I told Esmé we must ring the pigs' noses to stop them rooting. She grudged the cost of rings, and learned of a cheaper method from some disreputable acquaintance. This entailed her begging a length of stout copper wire from a telegraph linesman. The wire was obtained without much difficulty, but in an atmosphere of hushed secrecy. We were then forced to cut the wire into short pieces. We sharpened one end with a file, and thrust it through the upper edge of the snout, and with a pair of pliers twisted the two ends of the wire together over a stick. When the stick was withdrawn a perfect ring was left. Esmé declined to be present at the operation, but was pleased with the result.

Each day she boiled her swill in a disused copper in an old washhouse. It is illegal to feed unboiled swill in case microbes start any of the highly infectious pig epidemics. And to the stew she would sometimes add armfuls of nettles gathered from around the buildings. The pigs seemed to appreciate her care, for they thrived.

In four months' time she took the car and fetched a pig dealer whom she had met somewhere or other. She gave him lunch and a bottle of beer. Afterwards the two of them sat on a wall and watched the twenty-nine pampered pigs as they roamed about behind the house. The reluctant dealer was kept there all afternoon, but that evening Esmé showed me his cheque and her accounts. She had made over thirty pounds net profit.

At once I made up my mind to buy a hundred pigs the following spring. But I thought it wise to do a little dealing in pigs during the winter, in order to gain a first-hand knowledge of types and prices. Dealing is the hardest of schools, and all mistakes are so costly that one takes care not to repeat them.

One cold winter day Esmé and I drove about fifteen miles

to a high, bleak plateau on which stands the village of Cerrig-y-Drudion, which means 'the Stones of the Druids'. The farmers around there are great breeders of the hardy Welsh black cattle, and most of them keep a few pigs to dispose of their buttermilk. I had sold the baby Austin and bought a larger car in order to accommodate the pigs I was hoping to buy through the winter, for I intended to buy large pigs ready for killing, and to sell them within a day or two. I have a method of finding stock when I wish to buy. I drive to the largest farm I can find, and if I have been unable to see a name written on the gate or on a signboard or cart, ask to see 'the boss'. When he appears I say that Mr Jones at Llanrwst fair told me that he, 'the boss', wished to sell a pig or a cow, according to what stock I require.

Thus Esmé and I turned in at the gate of a big farm near Cerrig-y-Drudion and drew up in the yard. A lank, secretive youth of about seventeen peered a moment out of the dark doorway of a building, and melted back into the gloom. As I jumped out of the car two friendly sheepdogs trotted up, but turned from me to Luck. Luck always insists on coming in the car and on venting his truculence during these chance encounters. I left the dogs walking round one another stiff-legged, their tempers on a hair-trigger. In the building the youth was helping an older man with a cow. The man was obviously the boy's father. He was tallish and lean, with a limp black moustache very dark against his yellow skin. They were about to dose the cow. The youth had his arm over the cow's neck, grasping a horn, and his other hand was twisting up the cow's head by a firm grip on the nose, a finger in each nostril. His father dexterously inserted the tip of a hollow horn in the animal's mouth, and poured down a pink mixture which smelled of aniseed. The cow gurgled, and when released coughed a little, with the impersonal passivity which raises a cow so far above the frailties of human passion.

The man said good-day to me in Welsh, and the boy sat on a box, lit a cigarette, and became a spectator. I had by this time acquired a certain notoriety in my little world, I suppose by my incongruity – and Esmé's. The farmer knew perfectly well who I was, but, coming of a subdued race, he had that mentality which allowed him to give nothing away. I returned his greeting in English and inquired after the cow.

'She have the hoosh,' he said.

The hoosh is an ailment known as 'husk'. Long worms, as fine as threads of silk, gather in a cow's throat, and set her coughing to such an extent that she soon loses condition.

'I have give her a red drench,' added the farmer, and he produced the empty bottle for inspection. The famous Red Drink cured hoven, husk, garget and all udder ills, restored lost appetite, and was a fine tonic before and after calving. The cow showed no immediate ill-effects. With a fine, careless confidence I offered to cure the cow at once with some turpentine and linseed oil, mixed equally. This offer so warmed the farmer's heart that he jerked his head towards the car and whispered reverently, 'Be that the missis? I have want to meet your missis.'

I called to Esmé, and she came into the building wearing her detached smile. The farmer seized her hand and held it. Welshmen rarely shake a hand.

'I have seen you about,' he said archly, 'but you have not notice me. I hear much things about the master and you. Anwyl! Well! Well! Well!'

He went into fits of laughter, retaining his grip on Esmé's hand, and winking at me as if to say that he too could be obscene, lawless, criminal, mad or drunk. Neither Esmé nor I knew in which category we were, and smiled back at him as inscrutably as possible. I reminded him about his cow and the hoosh. He produced the ingredients which I wanted from some dirty bottles on a cobwebby shelf. We mixed the linseed and the turpentine in a jam-jar, and the boy went to the house to fetch me a teaspoon. When he returned we tilted up the cow's head again, and I poured a teaspoonful of the liquid into each nostril. I had no notion what the effect would be, and was startled when the cow broke away from the boy and sank to the ground with a gasp. The two men were delighted with this strong reaction, and looked from me to the heaving beast with a faith which would have flattered me had I not expected the cow to die. I calculated the value of the animal, and reckoned I should have to pay about twenty pounds. I avoided Esmé's eye.

And suddenly the cow lurched to her feet, coughed several times vigorously, and began to chew her cud.

'It always works,' I told the two disciples, who were eyeing me as if I were a witch-doctor. 'Repeat that daily for a few days, and you'll find the fumes will kill the worms.'

The cow was cured within a short time, and the farmer repaid me by sidling up to me at sales with titbits of inside information which were often of the greatest value. When I first went to Dyffryn I was a foreigner, a newcomer to the district, and a novice who knew absolutely nothing about farming, and so I was at a great disadvantage in not having knowledge of my own nor access to the knowledge of others. Thus I made a point of cultivating reliable informants who kept me posted in the intrigues and rogueries of my world.

After the healing there was a movement towards the house.

I felt that had the farmer been a High Churchman he would have crossed himself. Esmé and I were ushered through a big stone-floored dairy into a dark, cool kitchen. Slate slabs formed shelves around the dairy, and on them stood baskets of eggs, rows of neat pounds of butter with the form of a cow moulded on their tops, and many brown earthenware crocks of milk standing for the churn. The kitchen was in the pattern of thousands of Welsh farm kitchens. The walls were papered with a yellow-brown flowered design. The beamed ceiling was whitewashed, and the oak beams were hidden under the flaking lime. Hams hung from hooks in the beams. The flagged floor was newly washed, and a strip of coco matting lay on the hearth. A smug cat whose colouring matched the walls lay on the mat. A dog stood on the threshold regarding the cat, but knowing that he must not pass beyond the dairy. There was a dresser laden with willow-pattern plates and pewter, a grandfather clock whose dial showed also the phases of the moon, an oak dower-chest, carved 'E.M.1695', of dubious antiquity, and on the mantel was a row of spotless shining brasses and jam skellets. Propped against an ancient brass dog-collar was a postcard with a flowery birthday design addressed to Mrs Roberts, Bryntirion.

A stout, cheerful woman greeted us, wiping her hands on her sack apron before shaking ours. She had been warned of our arrival when the boy went for the teaspoon.

'Good morning, Mrs Roberts,' I said, 'we've just been killing one of your husband's cows.'

Esmé gave a mirthless laugh as we were led to a table set with homemade bread and butter, cold mutton, tinned apricots and tea. Roberts ate with us, while his wife urgently pressed fresh supplies on her guests. It was difficult to refuse, as she seemed always to choose times when our mouths were full. Presently I said to our host that I had been told he had some pigs to sell. He had been intensely curious as to our purpose in visiting him, but would have bitten his tongue off rather than ask. The frown of concentration left his face at once.

'I have prezactly what you like!' he exclaimed. He leaped up from his bench, seized his hat and stick, and clattered out in front of us to a shed. He opened the door gingerly and showed us two very nice bacon pigs, which weighed about eleven score. Pigs are always estimated by the score of pounds dead-weight. An expert can guess to within a pound or two, but my guess was, of course, very approximate. At once I began to work out how much the pigs were worth. Esmé's pamphlets had told me that a pig's dressed carcass was about sixty per cent of the live weight. Each of these pigs, then, would kill out at about seven score. Bacon was at that time quoted at thirteen shillings a score, so that a butcher would pay about ninety shillings apiece for the pigs. I did not want large profit so much as experience, and I decided to offer eighty shillings. This would leave me a margin for carriage, a few days' food, and the auctioneer's selling commission. I should be content to break square on the deal.

'Grand pigs!' said Roberts, with a knowing nod. 'Beautiful!'

I agreed. 'Pity they're too big for best quality. The butchers insist on having them about ten score.'

'Them is just right, mister. You don't want them no smaller. I have give tiptop food.'

Slung over the beams of the roof were several Indian meal-bags.

'I can see you've done them well, Mr Roberts, but if a butcher knew they'd been fed on Indian meal he'd say the flesh was tainted. It does make the fat yellow – although it's a temptation to use it when it's as cheap as it is today.'

'Anwyl! It's no' cheap when the bills is to pay. I won't have nothing in them pigs.'

'Well, how much do you expect to get?' 'I must have four pun, isn't it?'

Farmers sometimes are out of touch with the market. When this is so they usually cover themselves by making certain of asking too much. Occasionally, as now, they ask too little. I could follow Roberts' economics. He would charge nothing for labour, since it cost him nothing out of pocket for himself and his son. Rent would be too insignificant for him to notice, and interest on capital would not occur to him. The pigs were two from a litter of eight borne by his sow. Six he had sold at eight weeks for twenty-five shillings each. But he would never cost in at twenty-five shillings the two which he had kept. And with the help of Esmé's literature I knew what his food bill would be.

The pigs weighed two hundred and twenty pounds each. As weaners they would have weighed thirty. Therefore they had gained a hundred and ninety pounds. On a general farm it needs four pounds of food to add one pound of flesh to a pig. So each pig had consumed four times a hundred and ninety pounds, or about seven hundredweights, of meal. At seven shillings and six-pence a hundredweight the food bill would amount to about fifty-two shillings a pig. In asking eighty shillings Roberts was calculating a false profit of twenty-eight shillings each. Yet he could have had nearly that for the pigs as weaners.

'Eighty is a bit high, Mr Roberts. Remember that you are selling at home. There will be no carriage or selling commission, and you won't have to waste a day at the fair.'

'I must have four pun, indeed!'

'They're nice pigs, I know. I'll give you seventy-two-and-six.'

'By damn, no!'

'You must meet me a bit!'

'Well, they be on sale. You must pay four pun, and I'll give you five shillings' luck in them.'

Luck money helps to clinch many a bargain. Farmers are stiff-necked, and do not like to budge from their price, but the buyer can be met by a gift of luck money. A farmer may want fifteen pounds for a cow. The buyer may refuse to pay more than fourteen pounds. Then the farmer offers a sovereign luck money and receives a cheque for his price of fifteen pounds, which he proudly shows to his neighbours as proof of the quality of his beasts.

'Five shillings each!'

'Anwyl, no! There's no sense in it!'

'Well, we'll have to split. You can give me seven-and-six.'

At that I wrote him a cheque for eight pounds, which he accepted with a show of reluctance, and we struck hands to seal the bargain. He went into the house, and presently returned grumbling with three half-crowns.

Esmé and I had some other business to see to a few miles away, and we decided to collect the two pigs on the way home. When at last we returned to Roberts' yard it was growing dark, and was blowing up for a rainy, stormy night. I put up the hood and side-screens of the car, backed close up to the pig-shed, and opened the door of the tonneau. Roberts and the lad came out with a flickering stable lantern, and we went in to the pigs. Now that the time had come to load them we began to wonder how it could be done. The pigs were huge and, like all their breed, phenomenally strong. One of them made a rush at me, and I had to jab it heartily on the nose with my boot.

'They be a bit wild,' said Roberts rather anxiously. 'It will be more best to send a truck.'

But there was no room for both profit and truck. Pigs are susceptible to handling. If, say, a boar is treated as the wild beast which he can be, he becomes a dangerous creature. If he is scratched on the back, talked to and petted, he becomes friendly and docile. Roberts' pigs had never been petted. They were racing madly about the shed, uttering sharp grunts which sounded more like barks. We did once manage to lay hold of one by ears and tail, but it broke away. At last the lad made a running noose on a cord, and as a pig ran open-mouthed at him he caught the upper jaw in the noose. We led the brute outside. The lad climbed right through the back of the car and tugged the pig after him, while we assisted from the rear. Speedily we captured the second pig and induced him to enter.

By this time the first pig had investigated his prison. He decided to stick his head through the celluloid side-screens. It seemed as if he would force his way out, so we tied sacks over the outside of the side-screens. Roberts gave us a dubious good-bye, and we drove away. Almost at once there was trouble. I felt the seat of the car heave, and a slavering snout breathed hot on my cheek. Esmé gave a stifled squeak, and hit the pig on the nose as it was about to slide between us. Luck barked furiously and dived down under the scuttle, where he remained for the rest of the journey, growling thunderously.

I stopped the car and pulled out the handle of the jack. Esmé knelt on the seat facing the blackness of the rear compartment, and used the jack-handle to repel the continuous attempts of the pigs to get at us or with us. We were not sure which their intentions were. I drove as if a revolver were at my back. Every now and again a blast of hot breath struck the back of my neck, and Esmé battled in a grim silence.

Suddenly there was a violent lurch, and the car slewed across the road. I jumped out, thinking a spring had broken with the weight. The rain was whirling fantastically in the light of the car's lamps as the wind became its partner in a dance. From the front seat came exclamations from Esmé as she repelled an advance. I found that the weight of the pigs had deflected the spring so far that the side of the body had wedged against the side of the spring and the mudguard was binding the wheel. It seemed possible to free the spring by rocking the car. The pigs showed a determination to stand on the weak side, and it took ten minutes to dissuade them. It was like deliberately baiting a pair of furious but temporarily quiescent lions. At last we moved the brutes over, I heaved on the car, and the spring came free with a jerk.

The spring jammed four more times before we reached home.

Two days later I sold the pigs at a market in Anglesey. They showed a profit of fifteen shillings each. We speedily found that by going into remote districts we were able to buy pigs well below their market value. Very soon this sort of dealing began to seem immoral to us, and as we began to learn exact market values we were able to offer the sellers a price which was fair to both parties. Unfortunately this price was sometimes more than that asked. Therefore we realised that we should quickly gain a reputation for foolishness rather than honesty. But by then my education was sufficiently advanced for me to give up my excursions and to await the spring equipped with new knowledge.

On the whole I think that the dealer is a parasite. It is true that he acts as a catalyst between buyer and seller, but if he is to make a living he must either buy too cheaply or sell too dear, so that his profit is made at the expense of the producer or the consumer. The dealer is not a necessary agent, because the auction marts fulfil all his functions more efficiently and at far less cost to the industry. But farmers are lazy, and prefer often to sell at home. Then there is grave danger that the sharp-witted dealer will talk the vendor into selling below his price. The dealer can thus take his profits in the open market, or he can gain doubly by using his persuasive tongue to sell the stock privately above market prices. A good commercial traveller could sell refrigerators to Eskimos, so it is quite to be expected that such a specialist as a dealer should be able to outwit farmers.

Too much is expected of the farmer. He is a judge of many kinds of stock, he is a veterinary surgeon, a botanist, a chemist, an engineer, an architect, a surveyor, a foreman, a meteorologist, a buyer, a vendor and an advertising manager. And the only job in which he really fails is the last.

There are many honest dealers who seem to offer a fair price and to ask a fair price. But these are mostly big men who can afford to take a small individual profit on each of their innumerable deals. But the fact remains that they are quite unproductive, and must make their competence at the expense of the industry. Our twentieth-century economic structure has grown top-heavy with non-producing executives, and will inevitably collapse under the weight. And in farming particularly there is a host of small men who bleed the industry. They live by quick wits and by taking advantage of ignorance and credulity. Often they stoop to sharp practice, working in pairs like timber wolves. The first dealer goes to a farm and offers twelve pounds apiece for some bullocks. The beasts are worth more, and the farmer will not sell. A few days later the second dealer calls; but he tells specious tales of a slump and can offer only eleven pounds, so that when the first dealer calls back to renew his offer of twelve pounds the misled farmer is glad to sell, and to give heavy luck money into the bargain.

We were glad when our dealing was over. It left a nasty taste in our mouths.

10

The Dyffryn Pigs

THE REMAINDER OF THE WINTER PASSED, our days pigless, until the late mountain spring softened the severity of the landscape colouring. Esmé and I had learned that a good type of pig, a Large White–Welsh cross, was bred about thirty miles away on the Lleyn Peninsula. At Pwllheli every Wednesday was held a fair. A fair is quite different from an auction, for there is no auctioneer, and buyers and sellers work out their own damnation. One day we drove down to this fair. A great number of carts, car-trailers and even packing-cases were ranged around the big market square. Inns stood at convenient intervals behind the carts. Each vehicle was crammed with weaning pigs who slept with one eye open, burrowed under heaps of straw or bracken. Every now and again owners reached under the nets which covered the carts and stirred the inmates to their feet, in order to demonstrate their beauties to buyers who pretended to be disinterested.

Esmé and I began to work around the square in opposite directions, inquiring the price of any litters which appealed to us. At first the vendors thought that we were mentally unstable summer visitors who wanted pigs for pets. The prices at once became grossly inflated, until Esmé read several surprised old men a sharp lecture. She gave them offhand the price of weaners at Birmingham, Rugby and Carlisle, and suggested that the offending farmers should learn something of market values before they came to waste their time and hers. Esmé continued her inspection, rousing each litter till she could give a thorough examination, and until the square echoed with the cries of Bedlam.

The excitement became intense. A Welshman is full of curiosity and must solve any mystery. There was obviously something queer for them here. Perhaps someone was going to be done. Men ran into the inns to drag out acquaintances, and crowds followed us from cart to cart. Esmé's gallery became frantic with excitement, and the more agile kept scrambling ahead to secure points of vantage, like spectators at a golf match. The vendors were nudged and prompted to ask leading questions, to which Esmé, who was quite unconcerned, gave irrelevant answers. But at last a stir at the back of the crowd betokened the arrival of a personage. A hook-nosed, white-haired old man was thrust along the lane which opened respectfully through the mob. He was a sage of some repute. He walked slowly round Esmé like a horse-coper studying a mare, his lips pursed judicially. The crowd, now silent, watched him anxiously, hanging on his every movement. The old man stood still, his eyes fixed on an astral vision, and only his nodding head showed life as he listened to Esmé's remarks.

And presently he turned, calm and majestic among the undignified and excited disciples.

'That will be him from Dyffryn!' he intoned. 'And her is the Dyffryn's lady.'

I forgave the curiosity because of the title. Pwllheli was a satisfactory market. Weaners are usually sold at eight weeks, but in the Lleyn district it was customary to sell at eleven weeks. The price for the older pig seemed no more than was charged elsewhere for the younger, so we felt that we were buying three weeks' growth for nothing. We made weekly trips, and soon had a herd of a hundred and seven pigs, which was large enough.

I turned all the pigs into an enclosure of about thirty acres which was immediately behind the men's cottages. The enclosure was on a steep slope, and at the highest point stood a large stone-built and slated shed. As each successive purchase of pigs was brought to Dyffryn we drove them into the shed, where they could see for themselves the warmth and comfort to be found in the bed of bracken prepared for them. Pigs are hard to drive in a particular direction. Indeed, the best method seems to be to urge them the opposite way. When they have grown to know a place they can be lured by means of a bucket.

At night Esmé and I used often to stroll along to the shed and shine a torch on the pigs. The whole hundred and seven always lay fanwise in a corner. The great thing was to be the bottom pig. Artful pigs would go to bed early, jamming themselves tight in the corner, so that later pigs would pack in behind and on top, until a pyramid had risen over the first ones. The late-comers always took a flying leap into the mound and attempted to burrow deep into the mass of bodies, until their querulous neighbours bit them sharply in the ear. In the torchlight scores of cute eyes were cocked at us, and a chorus of sulky grunts made a continuous plaint until we went.

Pigs are notoriously subject to intestinal worms, and I decided to dose each lot as they were bought. A ground-up root called Santonin is effective when sprinkled in the food, but costs nearly a shilling a dose. So I used a mixture of castor-oil and five per cent oil of chenopodium. Great care has to be taken in the dosing, because the pig has a peculiar gullet. The foodpipe and the windpipe are very close together, and if in its excitement the pig takes the dose into the windpipe instant suffocation follows. I once killed a six-score porker in a moment by careless dosing. The speed of death is curious, because one would think the animal could hold its breath for some seconds at the least. However, as long as the pig can chew at something he seems to produce the correct oscillations of the gullet to swallow safely, so we thrust a thick stick in the mouth in the same position as a horse's bit. In the middle of the stick was a hole through which we poured the worm dose. The hundred and seven were dealt with successfully, though noisily.

I did not wish to set aside much labour for my pigs. If Esmé had costed her labour given to the twenty-nine there would have been a greatly reduced profit. So swill was ruled out for food. Again, a mash of meal for so many would have meant much mixing of food and water night and morning, and we should have had to make a great number of troughs. I decided to have a meal ration compressed into cubes. The system worked splendidly. Twice a day Thomas carried a sack into the field and trailed a long line of cubes across the grass, so that there was ample feeding space for each pig. The nuts could not be bolted. It was necessary to chew each one, so the younger pigs had as good a chance as the older to fill their bellies. After their meal the herd would trot off to drink in one of the streams.

We did not ring the noses of the pigs. Pigs do better unringed, because they find roots and minerals in the ground and remain more contented. But also I wanted them to do the work of a plough in the rough enclosure. By the end of the summer they had ploughed admirably, and had spread their manure evenly over the whole area. During the following winter we spread lime over the ground, and in the spring we hand-sowed a nutritious grass mixture. The seeds took well, and gave us a notably improved pasture.

Quite soon the method of feeding began to give trouble to the many walkers who wander through our land in summer. The pigs had become used to seeing Thomas appear at meal-times with a sack slung over his shoulder. One day when I was on the road two men were coming off the mountain, and were walking through the pig enclosure. The bracken was high, and not a pig was visible as they dozed in the shade of the ferns. But pigs are alert and curious. Some outpost thrust a head through the bracken to look at the trespassers. He saw rucksacks on their backs, and crept quietly out of his nest to tiptoe after them in the hope that he would have the first course of his meal undisturbed. But his immediate neighbours were suspicious on principle, and stood up officiously to make sure that no advantage was being taken. Instantly they squealed with mortification and raced after the first pig and his quarry. The dead slopes in a moment became alive with inquiring heads. To the startled visitors it must have seemed that Birnam wood was again going to Dunsinane, for the countless pigs which had materialised from nowhere were heavily camouflaged with fronds of bracken. The two walkers thought that they, not their sacks, were the quarry, and flung themselves over the fence and into the road as the shrieking mob was brought up short by the wires.

From time to time patrols of pigs would discover weak points in the fencing. Then the herd would pour through into the road.

I became used to excited AA scouts panting up to the house to tell me that 'the Dyffryn pigs were out again'. On hot days the pigs liked to lie on the warm tarmac oblivious of the traffic hold-up, until their flanks became so pink with sunburn that they squealed with pain when they were jostled. One day a lorry passed up the road laden with bags of Indian corn. Some of the sacks at the rear had burst, and a thin trail of corn was laid along the highway. The pigs discovered this at once, and scrambled off after the lorry in an attenuated string, as if running a paper-chase. The summer traffic became hopelessly disorganised for miles, and the police made such strong representations that we had at last to spend time in making the fences quite pig-proof.

The pigs did well. We had no losses, and there were no weaklings. The natural outdoor life prevented illness. Towards autumn it became time to sell the strongest of the herd. The biggest pigs were now at pork weight of between six score and seven score. It might perhaps have paid to keep them on until they were at bacon weight of about ten and a half score, but rough weather was approaching, and we should have had to put the pigs indoors. To feed so many pigs efficiently and economically the buildings would have had to be adapted roughly in the style of the Scandinavian pig-house. So we began to sell.

There is a good weekly auction mart near the coast resorts of Llandudno and Colwyn Bay. Each week we put our maximum load of six porkers into the back of the Morris and took them to the sale. The weekly load weighed a third of a ton, and the car became as animated as the pigs on corners. One day as Esmé and I were about to set off with our load a man arrived to see me about some business, and I was unable to accompany her. John Davies and Thomas were up on the mountain, so Esmé set off alone undaunted. She was travelling through Bettws-y-Coed, when the cheers of the inhabitants made her look round. A pig had evidently leaned against the door handle. The door had opened, and the pig had shot out stern first, but had retained with his forelegs a grip on the running-board. In his effort to maintain this position his rear legs were moving so fast along the road that Esmé swears they were blurred. She reached out behind and held him by one ear until the car stopped. Whereupon the pig scrambled back thankfully, amid the salutations of his colleagues.

A few miles farther on the weight deflected the weak rear spring so that it stuck against the body. The mudguard jammed the wheel, and the car halted abruptly. Esmé jumped out and tried to rock the spring free, but the weight of pigs was too much for her. However, the car had stopped on the outskirts of a village and exactly opposite a public lavatory. Esmé opened the car door and shepherded the pigs into it. She divided her time between jerking at the spring and repelling sorties from the lavatory. The spring jumped loose at last, and just as Esmé was wondering how she would load the pigs again the six rushed out of the lavatory, followed by a man whom Esmé had evidently besieged.

The man bore no malice, and did not seem to think the occurrence unusual. He helped to load.

Esmé hurried on towards the mart. The town was crowded, for market day had coincided with fair day, and swarms of farmers' wives were selling pounds of butter, eggs and hens in baskets. Stalls were set up in the streets, where sweating men sold china and cheap linoleum. The spring jammed in the square. It was close on time for the auction to begin. Esmé disembarked her cargo and drove them at a run between, under and over the stalls and baskets. Business was suspended until she and her charges entered the sale yard.

We were expecting about sixty shillings each for our porkers. The sale began at an ordinary price level. It is perhaps untrue to say that most butchers form a ring at an auction. But it is true to say that there is a general understanding between them not to bid against friends. In a small mart this means that there is virtually a ring, since all the butchers know one another. But today a strange buyer appeared. Esmé put the pigs up for sale on the condition that the buyer could take two, four or six at the price. Under these conditions the buyer invariably picks the best two animals, and bids afresh on the second pair, which are not so good and should sell cheaper. Then a fresh bid is made for the remainder at a still lower level.

But when the bidding was opened by a local man, the stranger chimed in. At once the silent acquaintance of the original bidder began to raise the price. Sixty shillings was quickly reached and passed. The auctioneer, who knew perfectly well that the stranger was going to be taught a lesson, pretended to study the pigs.

'Grand porkers! Straight off a mountain! Trust a lady to turn out nice pigs! Handy money at sixty-five! Who says seven-six? Come along, sir! It's against you now! Don't lose this healthy pen for half a crown! Seven-six! Thank you, sir! A beautiful pen, and cheap at sixty-seven-and-six! Fill it up, gentlemen – fill it up! Seventy shillings now! Stir them up Mrs Firbank! These gentlemen can't see them! Was that a bid, sir? Thank you, Mr Williams, three pun ten! Three pun ten! I'll take a shilling now! Who says eleven? As nice a lot as ever I've seen! You could cash these pigs anywhere! Eleven! Seventy-one shillings! And a half there! Seventy-one-six! Against you again, sir!'

And the unfortunate stranger, thoroughly worked up, offered seventy-two shillings. His rivals felt it unsafe to raise him further, and the auctioneer dropped his hammer. But in his confusion the buyer took the six at the inflated price. He walked off through the tittering crowd mentally kicking himself. It is surprising how even well-balanced men allow themselves to be carried away in a bidding duel.

Once at my annual sheep sale I put up a pen of twenty yearlings. I had bought them only the day before for thirty shillings. Two stubborn old men ran them up to forty-eight before one retired, leaving the winner to his Pyrrhic victory.

And, of course, most auctioneers pull non-existent bids from the crowd. I once took a pen of wether lambs to a local mart. There was a very bad trade, but my lambs seemed to be selling well. The bidding was brisk and animated, and the pen was knocked down. The auctioneer leaned over and whispered to me that he had not had a single bid from first to last. He had expected that someone, encouraged by the apparently busy competition, would throw in a genuine offer. I know of a firm of auctioneers who sell a good deal of their own stock at their sales without disclosing its origin. Any eager buyer will be run up in the bidding, until the last squeak is squeezed out of him before the hammer falls. Sometimes the whisper circulates about the ownership of the stock, and the auctioneers are cleverly led on to make their last false bid. I saw this happen once, but the man on the rostrum was not put out.

'Where's that man gone?' he yelled. 'Why won't he stand to his bid: I never trust these fellows who shift about at the back of the crowd?'

My own system at an auction is to wait until the bidding is nearly done. By then I have spotted the genuine bidders, and the price has gone so far that it is dangerous for either spectators or auctioneers to run the figure higher in case their offer is not capped.

Esmé returned from her epic trip delighted, if tired. Thanks to the stranger at the mart, the six pigs had realised about twelve shillings a head more than we had expected. Eventually the net profit on the whole herd of a hundred and seven was seventy-five pounds. The weaners had cost a hundred pounds, and they had eaten eighteen tons of food at a cost of a hundred and ninety pounds. I had used a rather expensive proprietary ration, and had possibly fed the pigs a little too heavily. When pigs are running loose it is pointless to feed to the pitch where the animals put on fat, for their activity works the surplus weight away. I ought to have fed just enough to keep the stock lean and growing. But I was well satisfied with my seventy-five pounds' profit on the total outlay of two hundred and ninety pounds for about four months. I charged nothing for labour, which was negligible, and which was fully offset by the manurial residue left by the pigs in the torn ground.

The success of summer pigs led me to keep pigs through the following winter. For a few pounds I managed to convert two cowsheds into rough imitations of the Scandinavian type of pig-house. A Scandinavian house has a central feeding passage four feet wide. On either side are pens ten feet long and eight feet deep. The food troughs run along the passageway, and can be filled from outside the pens. Behind the pens, running the length of each long outside wall, are the two dunging passages, one for each row of pens. The pigs enter these through a doorway from each pen, and when the door is open it swings across the passage, isolating every pen to its own section. To clean the passage the men drive all pigs from the passage back to their pens and close the doors on them. The passageway is then continuous. The most important rule of pig-housing is to give good ventilation. Pig dung gives off strong ammonia fumes which catch one's throat and make the eyes smart.

My hasty conversion was not elaborate, but it served its main purpose, which was to ease labour. I bought a hundred and sixty pigs in lots of ten at weekly intervals. As I bought the last lot the first purchases were ready for sale. I kept the numbers constant at a hundred and sixty by replenishing each empty pen. But nothing went right.

The enclosure floors were of concrete, and needed deep bedding to keep the pigs free from rheumatism. We used an enormous quantity of bedding, and as it began to run short, rheumatism became prevalent. I countered by making sleeping platforms of old railway sleepers, which the pigs found warm and comfortable. But soon there was a bad outbreak of colds. We spent days eliminating draughts from the old buildings. But it was food trouble which finally brought disaster.

I had approached a very well-known proprietary firm for advice. This firm did not make a complete ration, but sold a certain very important ingredient whose presence was essential in every ration. I then ordered my meal from a local firm to be mixed with the proprietary people's ingredient. Pigs began to die. I could see no cause of death, and took one of the corpses to a vet. He discovered intestinal irritation. I wrote to my proprietary firm and asked them to analyse the ration, which they did. They reported it good. But more pigs died. I sent the bodies to an agricultural research station, together with a sample of the food. Again the report on the meal was favourable, but the pigs were said to have died from a stomach irritation.

I would have changed my ration, but I had bought my meal forward from the local firm, as I had feared a rise in the price of foodstuffs. I could not therefore forfeit the twenty tons of meal for which I had contracted. And in any case two analyses had praised the ration. I still do not know what caused the trouble. But thirty-seven pigs died, and none of the survivors thrived. I was fortunate to escape from the venture with the loss of only a few pounds.

When pigs are good they are very, very good. When they are bad they are horrid. They are the most likeable and individual of all domestic animals except the dog – I am no horseman. Their cunning and their maddening persistence are more than offset by their comedy. The pig wins goodwill even as he challenges one's bad temper.

Pig-keepers in this country have for many years been handicapped by Danish competition. The officials who regulate our export trade have allowed the industrialist to find a small market in Denmark at the expense of home agriculture. Danish farms are mostly very small, and are worked by the family. There is no labour bill. And, again, the Danish system of collective marketing has given an indirect subsidy to her bacon exports. For only the first-quality meat has been sent to Great Britain. It has been sold here at a competitive price because the second-grade bacon has been consumed in Denmark after being sold at a high price. This has allowed the better quality to be sold comparatively cheaply in this country.

Denmark has for years been selling to us twice as much as she has bought. She has bought from us about sixteen million pounds' worth of goods and has sold us thirty-three million pounds' worth. And the bulk of her favourable trade surplus has been spent in

Germany. There is a double tragic irony in this. But, though it may seem callous to say so, it must indeed be a very ill wind to blow no good to anyone. The forcible closing of Denmark as a source of supply to us will inevitably be of advantage to our own pig-keepers at home and in the Empire. For alternative sources of supply must be found, and surely British farmers will have enough initiative to snatch a victory from defeat.

At home we can enormously expand our production of bacon and its subsidiaries. A sow bears a litter when she is twelve months old, and subsequently every six months, and as she can be expected to wean about eight piglets after each farrowing, we can very quickly increase our pig population to meet our full needs. And pig-keeping is especially suited to Great Britain. We are short of space in this country relatively to our population, and pigs kept intensively require very little room. Indeed, in Scandinavia there are two-storied pig-houses. And with pigs, fairly heavy labour is required. In this country any absorption of productive labour is a benefit.

Now that those thirty-three million pounds are to be diverted from Danish pockets to, one hopes, British purses, ninety-nine per cent of the money will flow back to the industrial market to purchase the equipment of which our farms are so sorely in need.

11

The Poultry Tragedy

OUR FORTUNES AT DYRFFRYN ARE tied to our sheep. We have cattle and we have our speculative ventures in pigs, but a bad sheep year still means a bad financial year. I began to think of some means to broaden our interests and to make us less dependent on the sheep flock. Unfortunately I thought of poultry. The British poultry industry is riddled with disease. At laying trials mortality is sometimes forty per cent. I think that after the Great War too many people took a fancy to hens. Ex-servicemen devoted their gratuities to this peaceful pursuit of poultry-keeping. The new poor invested the remnants of their money in the pastoral occupation. The demand for chicks out-stripped the legitimate supply, and dealers began to breed from poor stock, and some from diseased stock. Pullet eggs were hatched, with a mounting loss of stamina in the progeny, and when disease became rife the fowls had no resistance.

This state of affairs persists today, and it is only recently that the Government has begun to exercise some sort of supervision over the hatcheries. Many hatcheries are sound concerns which do their utmost to provide reliable stock, but the poultry industry is unparalleled in the spate of mushroom breeders of little experience and fewer morals who trap the dreaming theorist into buying their unsound chicks. At the best there is only the most modest competence in egg-production or in the fattening of birds for table.

The small margin, which is again due to unfair foreign competition, does not allow of any setback. I suppose that poultry-farmers easily head the bankruptcy list.

I was well aware of these dangers when I began to consider keeping a large permanent flock of hens. But the land at Dyffryn was absolutely clean. There was no possibility of disease germs being in the ground, or of contact between Dyffryn birds and strange birds. I expected heavy losses to occur at the start because of our severe climate. But by constant selective breeding from the strongest of the survivors I thought that I had a chance to build up so vigorous a flock that it might well form the foundation of a new poultry industry. I knew that if birds became acclimatised at Dyffryn and finally thrived, then I could sell my hatching eggs or my day-old chicks to any part of the British Isles in the sure knowledge that the stock would consider their home mild. And the birds would have regained that resistance to disease which their species has now lost.

Certain agricultural officials were much interested in my idea, and promised their full support in having Dyffryn enrolled as an accredited hatchery as soon as the time was ripe. I discussed with the officials possible reliable sources from which to buy my initial stock. I had decided to buy in the first year a thousand Rhode Island Red day-old chicks. The Rhode is a hardy, dual-purpose bird. The hen is a good layer, and the cock fattens well. Since half of each hatch are cockerels it is important to find a breed where the male birds come on readily for table. But so unhealthy was the whole poultry industry that even impartial experts were many weeks before they hesitantly gave me the name of a semi-official source from which to buy sound chicks.

I had decided to make all my equipment. The eager beginner in poultry is beset by firms who offer a vast variety of equipment which they advertise as indispensable. I first made brooders. They were adapted copies of a proprietary make. A brooder is usually in two tiers. Each tier is about seven feet long and two and a half feet wide. The floor is of fine wire-netting, and food troughs block in the sides. There is a water trough at one end. Two-thirds of the roof is enclosed with netting, and the brooder itself is dropped into the vacant third. The brooder is a square frame which rests across the sides. Felt is stretched across the frame, and drooping from it is a fold of soft flannel lightly filled with down. This cushion-like appendage nearly touches the wire floor, and the chicks thrust under it as if under a hen's breast.

Double felt curtains cut by numerous slits hang down from the four sides of the frame. The brooder can take over a hundred chicks.

Immediately under the brooder is a compartment for paraffin lamps. I had to buy these lamps, and I put three under each brooder. Each lamp had two wicks. The chicks are kept for about a month in the brooder, and wick after wick is turned out, until finally there is no heat at all. The droppings from the chicks fall through the wire floor on to removable trays, which in my home-made brooders were of zinc sheeting. I made a couple of these two-tier erections, and they had a total capacity of four hundred chicks. Esmé wished to manage the proposed poultry section, and I was only too glad. A woman's brain is much more suitable than a man's to deal with the patient detail work which is necessary. She placed an order for three deliveries of four hundred chicks to be sent at monthly intervals. This entailed having the first batch clear of the brooders in one month, in order to make room for the second. It is possible to sex the chicks fairly accurately at four weeks, so I began to make alternative accommodation of two kinds. I made movable outdoor arks for the pullets, and indoor carry-on brooders for the cocks. The arks were constructed of weather-boarding, and were seven feet long and three feet wide. They were designed to take a hundred birds. The floors were of wire-netting, with zinc dropping-trays beneath. Inside food troughs ran the full length of the arks, and two ramps gave easy access to the outside world. I made the arks as draught-proof as possible, but we again had recourse to heating, which was progressively cut down as the chicks grew and hardened. The carry-on brooders for the cocks were the same size as the early brooders, but gave greater headroom. There was, however, no actual cloth brooder and no provision for heating, as these brooders were for indoor use only.

The chicks arrived by rail in hay-filled boxes of twenty-five.

Chicks do not require food for the first twenty-four hours of their life. They are born with a meal in their stomachs, and losses are almost unheard of during their first day. We took the cheeping load home, and placed a hundred of the yellow balls of fluff under the canopy of each brooder. The chicks had been travelling all night and were hungry. At once heads began to appear through the slits in the felt curtains at the sides next to the food troughs, and in a moment a tight-packed row of golden heads was pecking rhythmically at the meal. By nightfall a few bold spirits were creeping out from beneath the canopy, and were feeding from food troughs in the open. The water at the end of the wire run was much too far away for them, so we had to put a shallow supply close to the curtains. Day by day we moved this water further away, until at the end of a week even the most timid chicks were drinking at the proper trough and were eating along the full length of the food troughs.

As dusk fell on the first day the chicks who had pioneered the wire crept back under the dark brooder. The last head was withdrawn through the curtain. Before bedtime Esmé and I went to have a last look. We opened the lamp compartments and peered up through the wire floors into the brooders. The chicks were crouched in a huddled mass over the lamps, their toes curved firmly over the strands of the netting. Scores of bright round eyes stared unwinkingly at the torch light. All was well.

But it was not well next morning. Several dead chicks were lying in the food troughs, and one was dead in a shallow pan of water. The food troughs were protected by close vertical bars, and I had fixed a horizontal slat across them which could be raised week by week to give more room for the growing heads. With daylight the chicks had begun to feed, and some had forced their way through the bars to die of cold. At their age half an hour's lack of heat was fatal. The remedy was to alter the troughs by fixing the vertical bars closer together. The chick in the water was a mystery. There was a depth of only a quarter of an inch. He must have stepped into the pan and died of cold from the feet up. I wrote at once, kotowing to the proprietary people, for drinking fountains made from jam jars. After these arrived we had no further water losses, but three more chicks died before they came.

Then we peeped under the brooders. Two or three dead bodies were under each. The little plastic forms were trodden flat as paper into the wire. We sought advice, and were told that the chicks must have been cold. They had huddled together for warmth, and some had gone underfoot in the press. We turned our lamps higher and spread them out in order to draw a group to squat over each lamp. There were still losses. We put newspapers over the floor under the canopy to eliminate upward draughts, and finally we laid sacks over the felt top. These combined measures eliminated the trouble, although I always expected to find the birds roasted. I now believe that insufficient heat is at the bottom of many chick troubles. The birds may survive, but they probably contract a constitutional weakness to colds.

During the first month, while Esmé lavished attention on her chicks and held an inquest on every death to avoid repetition of the fatal cause, I completed the arks and the carry-on brooders. On the day before the second batch of chicks was to arrive we sorted our birds. The pure Rhode Island Red is not easy for the amateur to sex. The cock is told by slower development of wing-feathering, by a stronger embryo comb, and by sturdier legs. Personally I found it easier to go by the bolder male eye. Many cross-breeds are sex-linked, and can be separated after a day or so. The males have different, unmistakable colour-markings from the hens owing to Mendelian law. We put our suspected pullets out into the arks, and the cocks into the carry-on brooders. And then we thoroughly cleaned and scrubbed the chick brooders against the arrival of their new tenants. We had lost forty chicks out of our first delivery, but Esmé had profited so well by her experience that the losses in the subsequent lots were only one or two per cent.

Day by day the pullet chicks became more venturesome and wandered farther afield from the arks. Birds of prey are hereditary enemies of young fowl, and on a sunny day the sudden shadow of a hand stretched over the pullets would send them streaking for cover. Perhaps the bomber will have induced the same congenital reaction in future generations of children. The cocks in their wire brooders grew strong too. Theirs was a duller, inactive life, but they had known nothing better and were content. Their legs grew as fat and fleshy as their breasts.

Presently we found that constant squatting against the wire floor was producing blisters on their breasts. This was uncomfortable for the birds and militated against their sale. So I had to change my wire floors in the carry-ons for floors of wood slats. Sometimes the cocks fought. They were not yet fully feathered, and the skin showed in places, so that a wound was visible. This would excite outbreaks of cannibalism, which we learned to check by plastering Stockholm tar over the cuts.

Accommodation had to be found now for the pullets and

cocks in the arks and the carry-ons. Their housing was going to be required for the present inhabitants of the early brooders who were shortly to be turned out in favour of the third batch of day-old chicks. I designed laying-houses into which to move the pullets, and a fattening battery for the cockerels. I made the laying-houses very squat to preserve them from destruction by the wind. They were ark-shaped, and with an inside dimension of eight feet by six. They would hold fifty birds each. We were buying twelve hundred chicks. Owing to our inexperience I was expecting only a thousand to reach maturity, and half of these would be cocks, so that we required ten houses for our five hundred laying birds. I bought timber as cheaply as possible, and we began to mass-produce the huts in two lots of five. My design was rather elaborate, and provided for laying-boxes accessible from outside and for a meal hopper which could be filled from outside with a week's supply, and which the birds could reach from within or without. The floors were to be of wire, and the droppings were to fall straight on to the grass. The houses were going to be mounted on skids so that they could be towed to fresh ground by the tractor from time to time.

The men became surprisingly good carpenters of a rough, practical, Robinson Crusoe type – though Thomas, who is notoriously careless of himself, smashed finger after finger with his hammer. One day he called to me, 'Will you fetch this plank, if you please?'

I asked him to wait a minute or two. I finished my immediate task, had a chat with John Davies, and strolled over to Thomas. He was patiently standing, motionless. He was wearing rubber boots. He had jumped off a bench on to a board through which was sticking upward a two-inch nail. It needed John Davies and me to wrench the nail from the boot and foot.

'Ta very much,' said Thomas. Then: 'Them huts is beginning to look lovely.'

John Davies was more rough-and-ready than Thomas. He insisted on using an enormous hammer which had been made for driving fencing-poles into the ground. He said its employment gave him exercise. He called it a 'pudding-bumper'. For the cockerels in the carry-ons I made a fattening battery. This was a three-tiered erection with thirty-six cages, each to hold six birds. The slatted floors were removable for scrubbing, and dropping-trays were under each tier. Our equipment was now complete, and each grade was in use. The newest chicks were in the brooders, the second batch were in the arks and the carry-ons, and the earliest lot were in the laying-houses and fattening crates.

Soon the early cockerels were ready for market. Many of them weighed six pounds at fifteen weeks of age, and as they had been hatched in January they were ready for the Easter trade.

Esmé had located a stall-holder named Fenner at a local market, and had persuaded him to come to see the birds and to make an offer. The man came, but the propeller-shaft of his car broke in climbing our hill, and he abandoned the vehicle to arrive on foot. He was a tall, good-looking man who had gone to seed. His face was not shaved, his linen was dirty, and though it was morning he smelled of whisky. His voice was cultured, and I could see that he despised his own educated accent. He probably disliked it as a link with a world from which he had slipped.

Esmé took him to the fattening battery, and he handled many of the birds. He noticed at once the few birds with blistered breasts. Then I took him into the house for more whisky. Finally it became imperative to offer him lunch. It was pathetic to watch his social graces. His manners were excellent, if rusty. But he was torn between the desire to tell us his history and prove his better origin by his voice and manners and his desire to conceal his evident fall from a better station by disguising the manners. Altogether the result was chaotic and sad. Anyway, I knew from his loose face muscles that drink had kicked him tumbling down the ladder. After lunch he made an offer for the birds of one-and-three per pound live-weight. This was a very good price, and I accepted at once. But he made free delivery to his house a condition of sale. I agreed.

Then we went to look at his broken-down car. Obviously no break-down lorry could reach it where it stood, and I offered to back it down to the highway. He accepted the offer with an eagerness which I soon understood, for when the scotches were removed from behind the wheels the car leaped backward. The brakes were useless. I turned the vehicle up against the hillside, and it stopped poised and ready to turn over. We adjusted the brakes, and at last reached safety on the main road.

We had a Ford Eight saloon car at that time, with provision behind to attach a trailer. We fed our cockerels heavily next day to increase their weight, and began to load. The Ford was soon full, and birds quickly fluttered out of the rear compartment to perch on the seats and the steering-wheel. Their excitement loosened their bowels disastrously. Meanwhile we carried the rest of the birds to the trailer, which I had overnight converted temporarily to a two-decker. Esmé and I fought our way into the Ford and thrust the protesting birds back into the rear. Almost as soon as we started, cockerels began to force their way through the slatted sides of the trailer. They would volplane gracefully to the ground, and were usually in a dazed condition, so that we could catch them. But every time that we opened the car door more birds hopped out. And while we caught them, the trailer superimposed another quota. It became like a game of solitaire, with the added distraction that the over-fed birds, upset by the motion, were being very sick. The car was a shambles.

We did at last reach the buyer's house, worn out but with all our birds. The house was large, built of a glaring red brick, with dark, blind-looking, uncurtained windows. It was set among monkey-puzzle trees and laurel bushes. The man came out and helped us unload. We picked the cockerels up by the legs in bunches of five or six. We tied the legs of each lot together and hung them squawking on to the hook of a spring-balance. When all were weighed and safely stowed away in a battery of coops we totalled up the weights, and the man paid a cheque. As we drove away a pale-faced woman peered with a haunted air through a window. I do not know to this day what ghost weighed upon the place, but I was glad to leave. Neither Esmé nor I saw the man again. His stall has changed hands.

We sold the succeeding two batches of cockerels in smaller lots to hotels and poulterers. But as summer advanced the price became progressively poorer, and many of the late birds realised only eightpence per pound. I doubt whether table birds really pay the producer. Perhaps the smallholder makes a little profit by fattening, but he would never charge his labour against the birds. If he did he would find that he was earning a starvation wage. And table birds may pay where the farming is on a vast scale, when the small individual profit adds up to a reasonable sum. But on the large, efficient farms much of the work is done by pupils who, even if they do not pay a premium for their instruction, at least work for nothing.

I should never again attempt fattening save as a means of disposing of male birds. Of course, it is possible even with a breed which is not sex-linked to buy day-old chicks whose sex is guaranteed. This sexing was pioneered by Japanese in this country, possibly on account of their manual dexterity. And most large hatcheries still employ Japanese for the purpose, though our own people are slowly being taught. The sexer has to squeeze the anus until the undeveloped genital parts show. If this is clumsily done, damage to the chicks results, and I have heard it said that one so handled may never be quite right again. Some breeds are heavy layers, and the cocks are useless for table, so they are killed as soon as they are sexed.

In July our oldest pullets began to lay. A heavy breed, such as a Rhode, does not lay for six or seven months under our arduous mountain conditions. Almost at once we had an epidemic of egg-eating. I had not supplied enough grit for the hens to eat, and they lacked shell-forming materials in their bodies, so that the soft-shelled eggs broke as they were laid, and the birds were tempted to eat. A larger supply of limestone or oyster-shell grit put close to each hut soon hardened the shells, but did not break the habit. We countered by stretching sacking over the nestboxes, with holes pierced at intervals. The hens laid on the sacking, and the eggs immediately dropped through the hole into the straw-filled space beneath. Even then some hens attempted to reach through after the eggs. Finally we darkened the nest boxes by a hanging curtain, and the trouble stopped.

The marketing of the eggs was not easy. We found a few retail customers among hotels and cafés, and for a time supplied a large Ministry of Labour camp with thirty dozen a week. But many eggs had to be sold wholesale to shops or at market. At times the price dropped to eightpence a dozen as foreign imports flooded into the country. In the winter the price would rise to half a crown, but at that time of year the egg yield was low. Even so, I believe that we could have cleared our cocks at a small profit, and that our eggs would have paid steadily, if not very handsomely. And our real goal was to build up in our isolated fastness a virile breed whose health and stamina would finally burst upon the decadent industry like a bombshell. But the bombshell burst on Dyffryn.

I used to move the laying-houses regularly. Sometimes they were dragged at great peril to myself as tractor driver along the steep slopes below the house. Time and again the tractor tilted on the hillside so far that it reached a point of balance and I had to lean over to stop it rolling. And more than once when I was pulling uphill the slope and the torque combined to pick the front wheels off the ground. Often though, in dry weather, I ranged the houses neatly along the river on the bottom lands. One night when the huts were down in these meadows a sudden storm sprang up. It must have been midnight when the furious rattling of doors and windows woke me. The bed itself was shaking, and the house quivered as it braced itself against the wind. Rain thrashed against the open casement windows of the bedroom till I thought the glass would break.

'What about the hens?' asked Esmé. She has a great capacity for worrying at awkward times.

'Oh, they'll be quite all right!' said I.

The huts were in a place where there was no danger from floods, but the bottom lands are much exposed in a gale. However, the squat huts were only three feet six inches high, and their broad base was eight feet by six, while with their complement of hens they must have weighed seven hundredweights apiece. No wind could get sufficient grip on their design to move that weight. But I slept only fitfully just the same. At daybreak I peered from the bedroom window. Three hundred feet below, the rushing river was lapping the patch on which the huts were grouped, but I knew that it could rise little farther. As far as I could see, obscured as my sight was by the rain which whirled in slanting streaks along the valley, one or two of the huts had moved slightly under the buffeting. But that was all. Shivering, I made back for bed. And halfway there something struck me as being wrong. I looked again through the window. The huts were there still, in a tidy group. The river could not reach them. I stared, worried. And I realised suddenly what was puzzling me. There were eight huts there this morning. The night before there had been ten.

Esmé and I dressed in a moment, threw on oilskins, and fought our way down to the meadows. Except for a few scattered bits of timber, nothing was left of the missing two huts. The wind had hurled most of the débris into the river. Here and there, miraculously alive, we found hens crouched close to the ground, bedraggled and dazed. Most of them were in no sort of shelter, for a hen when disturbed at night is stupid with sleep and can scarcely move. We picked up the survivors and pushed them into the other huts. I struggled along to the cottages to tell Thomas what had happened. He was just getting up.

'Which huts be gone?' asked Thomas. 'The two nearest the river.'

'Well, there's luck for you. There ben't a dozen hens go in them two. I have mean to tell you they be crowding into the top ones.'

This was a great relief. When the storm died away that afternoon, Esmé and I let out the imprisoned birds and took a count. Not one hen had been lost. We congratulated ourselves heartily, in spite of the lost huts.

'That hen's damaged a bit, though,' I pointed out. A bird was moving rather drunkenly, a wing dragging on the ground. But when we caught her no injury was visible. Our hearts felt like lead. We had never yet seen fowl paralysis, but the symptoms as we had read of them were unmistakable. It was bitter to be cast down in the very moment of relief.

Research stations everywhere are working to trace the cause of and to produce a cure for fowl paralysis. The cause is still obscure, the cure non-existent. The disease is the most dangerous and costly in the industry. We both knew that our grandiose schemes were doomed. We did not wish to control our disease and then to try to compete with the big hatcheries, who had also got their trouble under control, though it was there nevertheless. We had wished to be above suspicion. We had wanted to sell our chicks as being clear of all contact with disease, and with all the stamina of healthy parentage behind them. Esmé and I knew that we were finished as poultry farmers. Possibly we could have built up our flock again from the birds who would survive the deaths which we knew were coming. But Dyffryn climate would have been a hopeless handicap for commercial egg production. Only by the sale of hardened breeding stock could we have turned our climate from a drawback to a selling point.

Fowl paralysis, they say, can remain latent in a strain of fowl through perhaps generations. Then a period of hardship may spur the disease to activity. There is no doubt that this is what happened in our case. There was no record of the disease at the source from which our chicks were bought, but the trouble spread so rapidly through our flock that the root must have been congenital. The birds went wrong in twos and threes each day. First the wings would droop, then the legs became cramped and useless, so that the fowls struggled about like game birds crippled with shot. They would linger for days and even weeks before dying. We saved a little money from the wreck by selling the birds to Ministry of Agriculture experimental stations for research and observation. They paid four shillings for every bird received, but that was little recompense, for the market price of the pullets healthy would have been twice as much. The great amount of labour and the material used in making the equipment were wasted utterly.

I had hoped, too, to use the hens to improve the land. Much of the Dyffryn lower land is too rocky and broken to permit of harrowing or manuring, but the poultry was beginning to do the work admirably. The moss was torn out of the ground, the fibrous surface mat of decaying vegetation was broken up and the soil underneath aerated, and rich poultry manure was evenly distributed. The hens would have been worth keeping at little profit if only for the improvement they were making in the pasture. But as far as we at Dyffryn are concerned, the poultry industry must still await its Messiah.

12

The Washing

EARLY JULY IS WASHING TIME around Dyffryn. The wool is rising on the backs of the sheep, and they will soon be fit to shear. But the grease and sweat of a twelvemonth are in the fleece, and must be washed out before shearing day. Wool buyers subtract as much as 25 per cent from the price if fleeces are unwashed, because the grease weighs heavy. And we want to gather, too, because we have not seen the flock as a whole since cutting day. Maggots may have developed, and the sufferers will need attention. And on the credit side we expect to find unmarked lambs to add to our count, for many sheep are inevitably missed when we gather the flock down to the ffridd and the meadows for lambing. These ewes will have brought their lambs in splendid isolation on the mountain, and there will be more to castrate and to earmark.

For several days John Davies and Thomas have been helping neighbours to gather and wash. And when our turn comes, all the twelve or thirteen men meet at Dyffryn house, for the gathering is not done in the usual routine. There are now close on three thousand head of sheep up on the mountain. The lambs are still young, and if this large flock were mixed together they would have great difficulty in finding their mothers again after the work was done. So we gather the mountain in two halves. We walk up through the centre and wheel eastward to collect the sheep for the morning's work. And when these are done with, and after we have lunched, we climb up again and sweep the west half of the mountain.

We go up through the ffridd to the mountain in two parties. The main group cuts across the centre of the mountain, and a smaller party bears to the right to block the east boundary. Presently the top men of each line work round until they see each other. A ripple of shouts echoes round the semicircle of men, and thin, distant barks and whistles startle the sheep. The line constricts slowly; the isolated couples merge into troops, the troops into a seething flock. When the men sight the mountain wall the sheep are in a huddled, bleating mass near the iron gate, and the lambs, lost and excited, are doing their utmost to break back through the ring of panting dogs. We do not take the flock down to the main pens, but drive it to a pen which is just below the mountain wall.

In a moment now there will be a respite, and the men can take time to chat. News in hill countries spreads slowly beyond the tight compartment of each valley, and at gatherings there are representatives from several valleys, so there is much to be said. Perhaps one of the men has become a father. Thomas makes crude play at him with the castrators, and John Davies offers to earmark the newly born baby. Or little Davydd Pentref will have had another of his comical mishaps. On one occasion he looked very sleepy.

'Well, wake up, Davydd Bach!' cried Thomas, throwing a lump of turf at him.

'Anwyl! Let me alone!' says Davydd, winking at me. 'I have been awake all night.'

'You be old enough to let your wife have a rest.'

'Well, damn! I was out in a bog, not in bed. I have go fetch a load of peat with the old mare, and it come dark on me. She went to a soft place and sink up to her belly, man. So I loose the sledge and shove a plank under her chin, while I run to fetch help. Well, there ben't no one in nowhere because of the Eisteddfod, so I go back by myself and dig her out all night.'

'What time have she go down?' asks Gwylim Pantffynnon. 'About nine o'clock,' says Davydd.

'Well, that be funny!' remarks Gwylim. 'I have see you leaving the taproom of the Siabod Hotel at ten, and that's three mile from Pentref.'

'It ben't no wonder he have slept in a bog then,' says Davies.

'Did you get the mare out in the end, Davydd?' I ask suspiciously.

'Aye, man! I have dig a tunnel under her and carry her out on my back.' And Davydd goes into fits of laughter until his eyes are streaming.

'No wonder she have jump out,' says Davies. 'She must have think you was a horse fly.' I suggest that we stop telling lies in order to do some work, and the men fling their raincoats over the drystone wall of the pen, take off their jackets, and stuff their sticks into the chinks between the stones. The dogs, led by Luck and Mot, race madly round and round the top of the wall, tumbling stones down on all and sundry.

In our part of the world we do not wash the lambs, because they will not be shorn. Some districts do shear their lambs. This is not so much to gain the wool, which is little enough, and too short in the staple to be valuable, but to ensure that the close-cropped fleece will not be caught in thorns and brambles when the lamb is sent away to winter. But as we do not want to wash our lambs we have to catch them from the flock and shut them into another pen. The men form a chain and toss the lambs from one to another towards their pen. They are growing heavy, and the work is hard. Now and again someone shouts, 'White lamb!' and the animal is diverted to a little pen of its own. A white lamb is the term for one which has not been earmarked, probably through having been born on the mountain. The lambs in the main pen decrease in number, until at last the men split up and search individually before the last struggling lamb is thrust through into the other pen.

At once John Davies and Thomas take shears and wander about among the ewes to catch and clip any whose soiled tails are likely to attract egg-laying blow fly. One or two ewes will have maggots on them. The crawling white patch spreads, if neglected, from the tail on to the back. Thomas and his father clip away all the wool from around the affected area, and from a bottle I pour on oil which is at once an unguent and a repellent to future attacks. We put these ewes in with the lambs so that they will not be washed. Meanwhile the other men catch strange sheep and put them into yet another pen. Sometimes there is one old ewe whose gap-toothed mouth proclaims great age. Someone has shamefacedly to ask whence she comes. The men crowd around. Some finger the ears, and others look at the pitchmark. Thomas, who is inquisitive and hates to be beaten by a problem, announces definitely that she belongs to a farm beyond Beddgelert, ten miles distant. The argument grows furious, for such discussion is meat and drink to the mountain Welsh, whose past, present and future has been, is, and will be sheep. Old Bob Owen, Henblas, does not approach on these occasions. A judge does not descend to the well of the court. In desperation Thomas appeals to him, for the old man is reputed to know every earmark within a twenty-mile radius.

'She do come from Pandy at Dolwyddelan,' he says quietly, and spits to close the subject.

Thomas will go away sulking, and the men continue their work. I put a mark on such a ewe, so that we can catch her out when the flock is driven past the buildings on the way to the washing place. I can then shut her up and send a postcard to her owner.

On inspection a few strange ewes will prove to have had lambs; some of the lambs will have been lost. A fox, a crow or a crack in the rocks has caused their end. The bereaved mothers' udders are swollen and hard, the milk is thick and cheesy. No lamb has sucked for days. But another strange ewe may have a soft, loose udder. Her milk will show clear and fresh. She needs a lamb. Her owner, or her owner's shepherd, will put her among the white lambs. We watch her over the wall. She stands a moment with an eye cocked at Luck, who is watching her officiously. I call him away. The ewe trots among the lambs, whickering. A lamb bleats suddenly and with a note of complaint. The two run together, and the lamb dives immediately under the ewe's belly to the udder and wriggles his tail ecstatically. The ewe turns her head to sniff at him, then swivels her eyes up as much as to say, 'It's all right. Here he is!'

The owner or his man leaps over the wall and earmarks the lamb for its farm. All the rest of the white lambs are mine now, and we quickly cut and earmark them. We open the gate of the main pen, and the foremost sheep regard suspiciously the way to liberty. Then one trots forward, hesitates and shies through the opening. In a moment the others follow, pouring like sand through the constriction of an hour-glass. Some ewes jump wildly across the threshold in fear of an ambush laid behind the wall. The dogs rush in front of the flock, crouching belly to earth and giving way in short rushes in their efforts to keep the sheep within bounds. We come to the bottom of the ffridd, cross the road and drive on across the meadows to the river. Each year John Davies runs up a wire pen with funnelled sides which narrow to a little stone jetty projecting into a deep, wide pool. The water is so clear that I can see the smooth oval blue stones which form the bottom of the pool eight feet down. They lie lapped one upon the other like fish-scales. A little shoal of rainbow trout flash away downstream from some resting place under the jetty.

The men pen the patient sheep. We rush two score of ewes down the funnel onto the jetty, and a line of men holds them tight-packed. The foremost ewes regard the water as though it were boiling pitch. Unfooted by the press, one or two tumble in, and at once strike strongly for the shelving shingle across the river. This is a pity, for unless the sheep are thrown by hand they do not properly submerge, but scramble out with dry backs. I wade through the tight-packed mass of bodies to the edge of the jetty and begin to pitch the sheep head first into the water. The men press harder and harder behind the ewes in an attempt to force me off the edge into the water. If they fail they at once resort to adding more sheep to the original bunch, so that my turn at pitching is prolonged until my arms ache. I am damned if I will give in and cry for a rest. Presently Thomas will take pity on me and will change places with me. Usually he pushes the sheep over until I tell him to throw them. All our day's work is done to the end of these few minutes by the river, and it is a tremendous waste of preliminary time and labour to scamp the actual washing.

At last two men take up the last ewe and heave her with a 'One, two, three' far into the stream. Two men call their dogs, and cross to fetch the dripping flock across by the bridge. While the rest of us go up to lunch they drive the sheep back up the ffridd and pen them, mixed again with their lambs. Esmé has cold mutton, hot potatoes and carrots for us. We quickly empty a huge tea-urn, eat a big helping of jam tart, and fill the remaining corners of our stomachs with bread and cheese. This is all most unwise, as we always realise when we begin to toil up to the mountain for the second time. As we pass through the gate in the mountain wall the two men who have been guarding the washed flock in the pens turn the sheep through behind us. They go down at once for food, and then Esmé takes them in the car to the west boundary, where they climb up over the heartbreaking stones and heather to act as stops.

Again the drive sweeps over the mountain. John Davies and I round the Glyder. Two hours later we make distant contact with the stops, and wheel downward towards a set of pens which lie right on the boundary three miles from the pens of the morning. Once more we begin to sort the lambs. There are sure to be many more unmarked here, for the land is wilder and rougher, and it is difficult to gather clean for lambing. Sometimes one lamb is limping, a broken leg dangling. A dog must have driven him roughly among the rocks. I set the bone, which is so soft and plastic that the break will knit quickly. John Davies smears the leg with tar, puts a splint each side, and binds it tightly with a rag bandage. We cannot keep the lamb down for observation because we do not know his mother, but later in the summer John Davies will spot him in the pens and say, 'This flamer have come right, indeed!'

Several ewes will be suffering from maggots again. When the wind is from the north this section of the mountain is airless, sheltered by the circling hills. The sheep spend the sultry days among waist-high bracken, where the warm, humid conditions encourage the blowfly. We clip and dress the sufferers. Perhaps we find a sick ewe. She moves in circles, stumbling a little. On one side of her head is a small lump. A little bladder of water has formed, and is pressing on the brain. We pen her by herself, and old Bob Owen, Henblas, promises to come up next day to try to cure her. He has great skill in tapping through the bone to draw off the fluid with a syringe. The lambs are sorted, the strange ewes have found their lambs, the dirty tails are clipped. Then Thomas's stomach chronometer will prompt him to climb on to the wall and look towards the road a quarter of a mile away. Esmé is sure to be drawing up with the car.

'It do be tea time,' says Thomas, with satisfaction. He hops off the wall and trots over the tufted grass to help Esmé carry the basket of jam sandwiches and the can of tea. When he returns, his mouth is covered in jam, and the men crowd him away from the basket, while Esmé ladles out cups of strong sweet tea.

Old Bob Henblas has a very soft spot for Esmé, and when he can he takes her into a corner to make heavy gallantries. Two different worlds, different generations, upbringings and languages meet. The couple are firm friends. Most of the tired men sit along the walls and watch the phlegmatic sheep. Arguments wax on types and quality. Thomas always contradicts everyone, and his father becomes disgusted with him. To avoid trouble I spur everyone to his feet, and we drive the flock out of the pens, over the moor, across the road and into an enclosure by the river. This is three miles upstream from the washing place of the morning, and the flow of water is smaller. But John Davies has found time to build a dam across some shallows below a pool, so that there is a fair depth of water. More than once the press of sheep has broken down the temporary fencing. John Davies will notice me looking at his work anxiously.

'I have drove them posts firm with the pudding-bumper,' he assures me.

This second washing place is close to the road. Many cars stop whenever we are at work there, and excited visitors run across with cameras. When one of them asks what is happening, someone answers, 'We be teaching them to swim.'

Once a visitor asked how many head were penned, and was told ten thousand. He was impressed, and wished to see the owner. Old Owen Henblas pointed to little Davydd, who is notoriously unthrifty and lives by some mysterious process known only to himself. Davydd posed for a photograph, and to everyone's chagrin accepted a cigar. And when we have done we retrieve the wet flock. Thomas runs ahead to loose the lambs, and the grey stream flows towards the white until they mix inextricably. A babble of calls rises up, and suddenly a couple trots ahead and moves steadily up the mountain. The lamb tries to suck, but the ewe will not wait so near the scene of her trial. Two more couples follow, and again others in a never-ending stream. Some of the men live close to this west boundary, and tramp away with their dogs amid a chorus of good-nights. The rest of us pile into the car, a solid mass of men and dogs. The dogs snarl at one another and attempt to fight. Driving is difficult. We drop John Davies and Thomas at the cottages, and carry the rest of the men on to Capel Curig village, where they leave us and walk off on their respective ways through the thickening dusk. Esmé and I return, tired and content. Another day is well done.

13

The Shearing

SHEARING TIME IS UPON US at Dyffryn almost as soon as the washing is finished. The farms which washed early are ready to clip, for the wool buyers like the sheep to be shorn within ten days of washing. The short interval does not allow grease to accumulate again in the wool. We need forty men to shear the Dyffryn flock in one day, so I scarcely see John Davies and Thomas, who are away each day helping neighbours so that all the farms for miles round will send us help. In addition, I engage two or three extra men to go round to work on my behalf. Every day is filled, and it is hard to find the men to gather. Again we wish to handle the flock in two halves. We decide that our gatherers shall leave early from the farm where they are working on the eve of Dyffryn shearing. They will go straight from their shearing benches to their positions on our mountain, and we will gather one section before dark. The other part will be gathered at dawn. I am always in place early that evening. I start up towards the height of land, and reach the ridge in an hour and a half. Luck and Mot and I sit and look down the precipitous slopes into the next valley, where the Holyhead road lies like a careless grey thread. I smoke and watch Tryfan, whose sheer rock buttresses are in purple shadow as the sun sinks westward over the haze of Anglesey. Luck and Mot sit upright, prick-eared and expectant. I speak to them. They come to give a perfunctory lick at my hand, and resume at once their tense vigil. There is a primitive glory in the loneliness of a high place, and when tiny black dots intrude upon the landscape far below it is a sacrilege.

By seven o'clock the tired men are squatting a moment beside me. They will show signs of strain. For a fortnight they have been shearing from morning till night. They have been gathering with the daybreak, and have often gathered again before dusk. Some nights they have finished their shearing at dark, have had supper, and have walked over a mountain to be ready to gather on another farm at dawn. We spread out into our places, and are thankful when the tops are clear. Bad weather can make shearing time a nightmare. At Dyffryn, where the flock has to be gathered in two portions, a mist throws all our plans out of gear. If the night gathering fails we have to gather twice on shearing day. That means that a dozen men must be detached from the work to go up at midday for the remaining half of the flock. And sometimes rough weather will break on the shearing day. The flock speedily eats the food in the lower enclosures, and in the end we must turn it back to the mountain to feed, and must gather afresh when the weather seems to settle.

On the eve of shearing we drive the distant western half of the mountain, and as I round the Glyder I can see the dusk leaving England and slipping towards us over the Denbigh moors. On a clear evening the hogback of Siabod, high up, catches the last sun-rays which flame over the shoulders of Snowdon and her court. We hurry the sheep down into the great hollow of Cwmffynnon. We stumble wildly down the dangerous heather-covered hillside piled with loose rocks as we are chased by the hardening shadows. We reach at last the gentler slopes beside the road, and hurry the flock backward towards Dyffryn. Darkness smothers us even as we peer at half-light. I open a gate, and the men turn the flock through on to the road, where none of the sheep can become lost to us.

Thomas and I walk ahead to warn traffic. In the cars' headlights the countless eyes of the sheep twinkle like a bobbing cascade of emeralds. The fussing dogs' reflect ruby red. It is always pitch dark when we divert the flock into a field just beneath the windows of the house.

Esmé has a hot meal for us. At shearing time the kitchen is full of food. In a slow oven a huge round of beef is roasting against the hungry morrow. There are bowls and bowls of fruit jellies and custards. There are more jam tarts than the Knave of Hearts could ever have tackled. A stack of homebaked loaves is on the window-sill. Carrots, ready scraped, and peeled potatoes stand waiting in crocks of water. A big pan of porridge is slowly simmering for the gatherers' breakfast. It is often eleven o'clock by the time we have eaten, and in another four hours we must be on top of the Glyder once more to clear the remaining half of the mountain. Some of the men will vote for playing cards all night. But Thomas and Davies usually invite a few to snatch what sleep they can in the cottages, and Esmé and I provide some sort of sleeping quarters for the rest. We double lock the kitchen and its eatables against dogs and go to bed. I am too worried about the weather to sleep soundly, and from under the windows a tireless chorus of bleats beats about my ears like the sound of surf. The clamour of the alarum-clock finds me half awake at two in the morning. I stumble from bed, wash half-heartedly, and pull on my clothes, shivering in the keen breeze which plays through the window. I go round the house to rouse the men. Esmé, instantly awake and fresh, slips down to make tea and to cut bread and butter.

My first anxiety is always the weather, the ruler of Dyffryn.

Sometimes a moon silvers the valley. Sometimes heavy wet mist swirls about the house itself. And sometimes the sky is patchy, with white clouds lying like down against an inverted black carpet, while here and there groups of stars glimmer through the shifting gaps.

A ceaseless clamour arises from the sheep as they quest for mothers or children. And when the night is clear I hurry in, brimming with impatience to be gone before the weather can change.

We drink our tea in a morose silence. Afterwards the men fetch their dogs from where they are shut in dark buildings, and we begin to climb up the hillside dispiritedly. We stumble over stones or splash unseeing into streams, too weary to curse. Even the dogs labour quietly at heel. The mountain wall looms darker than the night, and we pass in silence through the iron gate. On upward we go, and as men begin to reach their places they sit down gloomily and extract comfort from tobacco, with their dogs couched beside them on the dewy turf.

John Davies and I pass three little depressed circles in the ground, the hollows just discernible as deeper shadows. It is an ancient gaer, a group of primitive huts. I feel that the dark little figures of the owners have slid down the thread of time and are stealthily keeping step with me. But Luck has never shown any sign, and a dog's judgment is infallible in such matters. I leave John Davies at last, and pick my way alone to the height of land. I sit where I had been before only a few hours earlier. But now I can only guess where lonely Tryfan stands in the mystery of the black abyss.

I know a place in Wales where a man may sit alone the night through if he dare take the vigil. At daybreak he is found stark mad, or else he will become a great poet. I want to pass a night there some day. Each time that I climb the Glyders to wait for light for the shearing gathering, I wonder why I do not come often to the high places in the darkness. For the velvet hand of the spaceless night soothes my jangled little human soul, and Luck and Mot, instant in perception, lay a head on either knee to share the balm. Perhaps I float awhile senseless on a dark sea of time, for when I look again towards Tryfan her jagged outline has become visible, like the serrated ridge on the back of a rearing dragon. The light has followed the dark in their ceaseless race around the world. Where not so long before the shadows had swept from England, treading on the heels of twilight, the wheel has turned full circle. The night draws in her skirts, and steps away westward like a dark enchantress as she avoids the warm embrace of day. The top of the Glyder behind me is touched with light. The summit becomes a flaming torch in a black sea. In a moment the golden glory of the dawn is on me and the sun lifts an orange face from behind Siabod. Its dull disc is part cloaked by grey-red cloud. The shades shrink down to the valleys, and the mountains lift their heads, rejoicing in the wake of day. Thirty miles away the windows of Harlech catch and throw back the pale sunbeams. On the shores of Cardigan Bay pools glimmer in the sand. Flat Anglesey springs to view like a floodlit map, far behind the cone of Tryfan.

Fancies flee shyly before the imperative hurry of work. I yell across the broken hillside. A thin shout comes back from John Davies. Luck scouts slinking along the ridge, while Mot dashes impetuously to link with Bett on the lower side. The whole line, unseen to me, sweeps forward joyfully in the good fortune of the crystal-clear air. By seven o'clock the sheep are safe in the main pens and we go to breakfast. Three long trestle tables are ready laid for a lunch for forty men. We gatherers sit at one table and eat porridge and egg and bacon. Usually as we are finishing, a few early shearers come in for a cup of tea. We never have much time to talk with them, and we leave them gossiping, and return to the pens to sort out the lambs and the strange ewes. John Davies goes off to make sure that the flock gathered on the previous night is secure, and then he harnesses Belle to the cart and fetches her to the pens, where we tie her up ready to carry the first load of wool.

Thomas sets out the long row of shearing benches, and I light a fire to heat the pitch-buckets. By eight o'clock a few shearers will be straggling down to the pens, and with them are sure to be a small boy or two, who are anxious to earn half a crown. One of the boys will be handed the 'strings', and also the picric acid with which to dab cuts. The strings are long strips of linen, traditionally torn from linen sugar-bags, with which each shearer ties the feet of the sheep on his bench. The men are never long in straddling their benches, for they are as anxious to finish the work as I am. They take their shears from their leather cases – for, like all good workmen, they are most careful with their tools – and hone them.

At Dyffryn the men must be worked hard if we are to finish before dark, and I always cut as much as possible the number of men who are on any job other than shearing. If the weather seems settled we do not even cart the wool away, and thus save two men who would be working with the cart. Davies takes on the job of lapping the wool as the fleeces fall one by one around the shearers' feet. One man is unable to keep up with the accumulating heaps, but at intervals the women come down with food, and they always clear up arrears. This job, too, leaves Davies free to roam about and to keep an eye on everything. Thomas has the task of stamping each new-shorn ewe with pitch, of untying her legs and tossing the strings to the boy. He is sure to scald himself before night. The only shearing day when he did not scald himself he was bitten for interfering in a dog-fight. I feel that I myself must do the hardest work in order to spur the other men, who are at their most competitive during shearing. But then I come to the task fresh, while they have been working at full pressure for a fortnight or more. Fortunately they never realise my advantage and their handicap. So I take my position in the sorting pen, where it falls to me to catch every sheep to be shorn, to carry it to the gate of the pen, and to hand it to the two men who supply the shearers. These men must be both robust and willing, for at times the sheep will be dealt with five or six a minute, and the farthest shearing bench is twenty-five yards from the pen.

Most years Emrys Jones, Castell Farm, and Price Pritchard, Garthllugydadaith, are at Dyffryn shearing. Garthllugydadaith means 'the farm which lies in the eye of the sun'. When these two are present I ask them to carry. Emrys Castell and Pricey Garth might be brothers. Both are short, sturdy and as full of energy as bouncing rubber balls. They have the same perpetual gap-toothed grin and the same fondness for practical jokes. They always accept the roving commission gladly. We pack the sorting pen tight with ewes so that I can reach sheep from the gate as they are immobilised in the tight-packed mass. I catch the first ewe and give her to Emrys. The long day has begun. Old Bob Henblas is always one of the early arrivals, and he invariably sets his bench in a sheltered angle next to the gate of the pens. Shearing day, congregating as it does thirty or forty men, is a social event of the greatest magnitude, so old Bob is sure to wear his best tweed suit (protected by overalls), his starched dicky, and his pearl-grey Homburg hat. But Emrys regards dignity as a bubble to be pricked, and will drop his first ewe suddenly on Bob's lap. If her thrashing legs kick off the old man's hat, so much the funnier. And Pricey Garth, not to be outdone as a comedian, will hurriedly think up a turn of his own. Maybe he will wink at me and indicate a yearling most of whose fleece has fallen off. And when I catch her for him he will give her to tiny Davydd Pentref, who is egg-bald. Everyone will guffaw and point out that like attracts like.

In spite of the skylarking the carriers rush out the sheep without pause. By the time the last man of the line has his first sheep, the men at the beginning have done and are yelling for more. There is no pause for any of us. Thomas, probably sucking a burned finger, starts to stamp the shorn sheep from his smoking bucket. The farm's ancient pitchmark, 'R', shows up well against the white, close-clipped wool. As he marks each ewe, Thomas swiftly unties her legs and tosses the string to the lad. The lad is racing continually up and down the line at the cry of 'String!' Now and then someone shouts 'Cut!' and the boy dabs picric acid on the wound. But there are few cuts. As far as possible I give the difficult sheep to Bob Henblas. Every year the wool rises from the skin of the sheep. It is the preliminary to sloughing the fleece. Some ewes would never actually lose the entire fleece, but the hereditary rise takes place just the same. Indeed, all animals, including birds and reptiles, have a looseness of the covering during summer. But on some sheep the wool rises late, and at shearing time it is still clinging close to the skin, and as the man's hand parts the fleece before the slicing shears, the skin is pulled up with the wool. Only a very skilled operator can then avoid making cuts. Bob Henblas clips first the short coarse wool under the belly before he gathers the four feet into a bunch and ties them deftly. Then he lays the ewe on her side on the bench between his thighs and slips the points of his shears under the curly wool of the neck. He works to and fro along the ribs, peeling off the wool between belly and backbone until he reaches the rump. The fleece hangs like a cloak doubled back over the spine, and one side of the sheep is white and naked. He turns the ewe over and begins again on the unclipped side. His shears, opening and closing in his tireless grip, seem to slide smoothly over the skin. There is no jerk in the movement and no hesitation. The whole fleece tumbles off the rump on to the ground, part held up by a few last strands. Old Owen snips through the remaining threads, clips the wool off either side of the broad tail, and dumps the clean-looking sheep on the ground.

From time to time more men arrive from distant farms, and by about nine o'clock there will be thirty shearing. Not all the men use benches. Some shear on their feet, bent low over the untied ewe. The position looks tiring, but the fastest shearer who ever comes to Dyffryn works this way, and keeps on all day. This youth's name is

Lloyd George Kitty Maud Jones, Ffridd-y-Foel. He is tall, black-haired, with a hard, wiry body. He was christened Lloyd George Jones, and would ordinarily have been known as Lloyd George Ffridd-y-Foel in order to distinguish him from his namesake, the politician, who has a house at Criccieth. But in this case there is a famous sister, who quite outshines the politician. Rumour has it that she would have defeated Messalina in open competition, even as Messalina put Scylla to shame. There are three more brothers, proud to be known by the sister's suffix. They are Vimy Ridge Kitty Maud Jones, Hill Sixty Kitty Maud Jones, and Granville Cadwallader Kitty Maud Jones.

As soon as my sorting pen becomes fairly empty the sheep have room to avoid me, and they become more tiring to catch. So now and then I rush to the main pen and drive in more, while Luck barks excitedly from the top of the wall. When I am away the carriers have to grab for themselves, and the shearers sometimes have to wait a moment for fresh ewes. Instantly they howl in unison at Emrys and Pricey, and abuse their negligence and idleness. But these two grin through their sweat and pick the most difficult sheep for the loudest shouters.

I return to my post, and together we catch up with the demand. On warm days I take off my shirt, and in the press of work even Emrys and Pricey become silent. There is no farming scene which is more busy-looking than shearing. The shearers bend over their sheep with tense attention. The boy rushes to and fro at the cry of 'Strings!' or 'Cut!' Thomas stamps the clean sheep from his smoking pitch-bucket, and his father laps the scattered fleeces into balls and tosses them towards the cart. The two carriers and I are streaming with sweat. And suddenly Esmé appears with some women helpers. The women pick their way carefully down the steep slope to the pens, carrying cans of tea and baskets of sandwiches. The shearers calculate their time nicely, and take care that Emrys and Pricey do not give them another ewe at the last moment. As each man finishes his sheep he stands up, mopping his face, and strolls over to where the food is on the grass. Old Bob Henblas will put down his ewe half finished, and remain seated. When Esmé carries tea over to him, the cunning old man will persuade her to take his place on the bench and to finish the sheep under his tuition.

But after the shortest break John Davies becomes restless. He fidgets about until one by one the men stub out their cigarettes and drift back to the benches. The carriers supply the line again, and, refreshed by the tea, the shearers work at a furious rate. Slowly the penned flock shrinks. There is usually two more hours' work on it, when Thomas checks his belly by somebody's watch and shouts that it is noon. Everyone hurriedly finishes his sheep, and we scramble up the hill to the old house. The scene might be a film set. As the rough men clatter to their benches Esmé doles out helpings at a side-table. Her self-conscious aides rush to and fro in a flurry of excitement, and giggle at the broad asides of the diners. And when the clink of cutlery dies away, and the beef and the tarts and the cheese have been washed down by pints of tea, Esmé scatters cigarettes on each table as a kind of bonus. After a moment's uproar there is quiet as each man smokes contentedly, with a spare tucked behind his ear. John Davies says suddenly, 'Well!'

The company rises like a well-drilled unit and clumps outside.

Before one o'clock the shearing is in full swing again. On shearing day I am haunted by fear of rain. Even one heavy shower can cause hopeless dislocation, for it is impossible to shear wet sheep, because the blades stick in the wool. And on fine days, when the wool is left to lie and the pile grows mountainous, my heart is in my mouth, because damp wool does not keep well, and the water in the fleeces leads to complications at the weighing. If I seek support from Davies he is very loath to take men away from the benches to cart wool.

'Carry on,' he always says, 'and chancet it.'

Sometimes in the midst of the activity I look at the western sky. Perhaps I will see white cumulus clouds with dark grey bellies drifting down the valley from Snowdon, like bladders floating on a blue sea. In sunshine the slanting rain streaks will show plainly like falling shafts of light. The men see the warning as quickly as I do, and always they bend yet more furiously to their work to cheat the rain of as many sheep as they can. The carriers become silent save for grunts, and I have scarce time to worry in the fury of the work. Men like Lloyd George Kitty Maud will be clipping their ewe in three minutes or so. If luck is with us, the clouds drift on to one side of us, soaking half the valley but leaving us bone-dry. At other times the Armada floats over us, sail on sail, to deluge us with her broadsides, and to leave us to reorganise our affairs as best we can.

During early afternoon the last sheep of the first flock come at last into the sorting pen. But the carriers are sure to tell anxious shearers that there are some two hundred yet to do. Presently the last ewe is carried out. As the men yell impatiently for more, the carriers grin derisively. By some clairvoyancy Esmé and her gang always appear at this moment like genies. They bring with them four huge cans of tea, but no food as yet. Davies and Thomas gulp theirs scalding hot and go off to fetch the second flock. The rest of us turn out the lambs to the shorn sheep in the enclosure, and two men volunteer to drive the reunited flock up the ffridd to the iron gate in the mountain wall. It is a relief to have a half of the flock out of the way. Then I set two men to carry fleeces to the building, and the wool-cart will ply backward and forward for the rest of the day.

We quickly pen the second flock when it arrives, and many hands help us speedily to catch out the lambs and the strangers. Again the carriers supply the line. But by now most people are tiring, and as the afternoon passes, the chatter dies away and there is little sound but the ceaseless grinding clip-clip of the shears. I tire too. Instead of ninety or a hundred pounds weight, each sheep grows, it seems, to be a couple of hundredweights. It becomes impossible to slip one arm under the neck, to grab the wool of the side with the other hand, and to whisk the animal to Emrys or Pricey. It begins to need a conscious effort to lift, and the constant struggles of each victim shorten one's temper. Hard corns form below the root of the nails on the first joint of each finger through twining in the harsh wool. And often, too, although I am physically very fit, cramp attacks the legs and arms. It seems as if the pens are filled with a limitless sea of grey backs, and it is fortunate that fatigue dulls misery, so that one can work like an automaton.

So broken does my spirit become that I forget on these after-noons to watch for the arrival of tea, and it is a never-failing sur-prise when cheers announce that Esmé and her myrmidons are down again with steaming cans and huge baskets of scones and sandwiches. The two men who have been taking the first flock up to the mountain invariably arrive back with the tea. Usually they are openly accused of lying in wait smoking in the bracken until they could see the women on the way. The respite is all too short. We start work once more, and the sheep roll out interminably. Two or three men have to leave, amid general execration, to perform duties on their own farms. But the diminishing shearers will only be the more on their mettle, and when the inevitable tea comes down again at six o'clock the second flock will be half done.

Frequently it is nearly nine when the last sheep is carried out, and everyone is too tired to joke about it. The cart will have caught up with the accumulation of wool, and as it goes off with the final fleeces the men plod stiffly up the hill to the house for supper. John Davies, Thomas and I hustle the sheep and lambs to the road and drive them up the valley, for this flock inhabits the west half of the mountain, and we want to turn them where they belong before dusk. It is always dark when I return to Dyffryn. Esmé will be sitting alone by the dying fire in the old house. The spirits have tiptoed from the dim corners to crowd about her. I sink down on the floor by her side, and we remain silent and motionless, too tired to speak or to move. But our minds are receptive, tuned to the infinite, because our bodies are too drugged by fatigue to intrude. Queer tricks of time play upon our stupid practical minds, and subconsciously we seek a key which just eludes us. The mystery of the past is not unlocked, but we both peer through the keyhole and see dimly the small, dark shapes who huddle round us for company. Esmé and I are doing the same work as they once did, and in their groping, introspective way they give us tongue-tied sympathy. The last glow of the fire dulls, and with the chill of midnight our companions go, irritated a little at our lack of full communion with them.

But Esmé and I have been up and doing for twenty-two hours at a stretch, so perhaps our fancies are a dream.

14

The Wool Sale

SHEARING CAN, OF COURSE, BE done mechanically. In Australia and New Zealand the vast flocks are shorn in long sheds with power-driven clippers run off a length of shafting. In this country many farms use electric shears. These are very like the clippers which a barber uses on the back of one's neck, and the tiny motor is contained in the actual head of the instrument, so that one has only to plug a length of cable into a lighting-point. But this makes the head rather heavy and unsuitable for continual work. The other type of shears is where several heads are worked by flexible cables off shafting which is run by an electric or petrol motor. Most English farms have flocks of a size which can be packed into a building, and no doubt the electric way is convenient, but at Dyffryn the portability of the self-contained petrol outfit appeals to me, and I may one day try it. These outfits consist of a small motor mounted on a trolley and working four heads. They cost about forty pounds. But the farm would not economically carry more than one set unless I could hire the machines out to neighbours. And people are conservative in Capel Curig. So we should be limited, I believe, to four shearing heads. I do not suppose that the mechanical way shears so very much quicker than the hand method. So although we could dispense with our army of men and with the expense of sending out help to repay, we should have to keep our flock

171

down for several days while we worked through it.

There is local prejudice against mechanical shearing because a local farmer used this method once and the sheep were so clean-clipped that many died of cold in a severe spell of weather which immediately followed the shearing. But now combs can be fittedto the clippers which allow almost any thickness of wool to be left on the sheep. And as the mechanical shears run over the sheep without leaving any ridges, slightly more wool will be taken off without leaving the ewe any colder.

I obtained an electric shears once for trial. John Davies is surprisingly open-minded about new-fangled methods, though a little clumsy. We tried the shears on about a hundred sheep without any trouble, except that Thomas gave himself an electric shock by feeling for the light-point with his finger when he was plugging in. The Welsh wool is coarse and wiry, and blunts the blades rather quickly, and the small, angular body of the sheep does not allow of making sweeps with the machine, so not much is gained by the modern way. And I think we should all miss our old-fashioned shearing day.

There is still some shearing to be done after the big day. Stray sheep come in from other farms, sometimes twenty or thirty at a time. And the rams have to be done. John Davies allows no one to touch these but himself. He drives the patriarchal flock of forty or so into a building and sets his bench on the roadside, while Thomas carries for him and stamps the shorn rams with pitch. Cars stop to watch, and Davies poses for photographs with a deprecating smile. He receives many cigarettes and half-crowns, and treats most cavalierly the laymen who question him. John Davies has a great faculty for introducing his photo into the daily papers. Once Esmé and I took a long winter holiday. One day we became snowbound in a hut in the Austrian Tyrol at six thousand feet. In a box I discovered an English daily picture paper which was a month old. On the front page was John Davies driving sheep with Bett through deep snow along the road in front of Dyffryn cottages. He looked quite ingenuous. In the old days when wethers were kept in the hills, the bulk of wool was large, and the price was reckoned to pay the rent of the farm. In my nine years of farming, the price I have received has fluctuated from threepence-halfpenny per pound to one-and-fourpence-three-farthings. In the years of the Great War Welsh wool was sold at three-and-six. Neighbours of mine refused that price, and held out for four shillings. Next year they had to clear it out at eightpence-halfpenny. Greediness never pays in farming. If a price seems too high to be economic, or if it shows a satisfactory profit, it is best to close the deal at once without holding out for more. A great number of people think that the sheep farmer makes his living from the wool. Yet these people would no doubt be ashamed if their ignorance was so profound on any subject but agriculture. At Dyffryn our wool clip usually averages between a hundred and fifty and two hundred pounds' worth. The bulk of the takings with sheep comes, of course, from the lambs. A Welsh fleece is rarely worth more than half a crown, while a store lamb is worth a sovereign.

Many shady people flutter like vultures over the sick carcass of British agriculture. But the greatest number of rogues batten on the wool trade. There are innumerable honest buyers too, and it is a pity that the activities of these straight men should often be suspect because of the crooked dealing of the rogues. Some dealers buy for themselves, sort the wool, and sell later to Bradford. Others act merely as agents for big firms, and take a commission from their principals of perhaps a farthing or a halfpenny per pound. And some Bradford firms send round their own representatives each year. The fleece is the main shield between the sheep and the stormy winter, and we deliberately breed our wool to be coarse and springy. Welsh wool is shot with a fibrous, wiry thread known as kemp, and it is this which renders the fleece resistant to cold and wet. But the kemp does not readily take dye, and remains white and bleached in the cloth. Thus our wool is used chiefly for rugs, blankets and carpets, though now and again ladies' fashions lean towards very rough tweeds. In those years Welsh wool sells at a higher price, and many a remote mountain bothy has extra winter luxuries because the vagaries of some leading dress designer have decreed hairy tweeds. One year Dyffryn wool was shipped to America, and Japan and Germany have occasionally been bidders in the Welsh market.

I always sell my wool to local dealers, though twice I have nearly lost my money. One year my regular dealer offered a very good price, but said that he could not pay at once. He offered a cheque post-dated a fortnight. This dealer had an order from a small mill in mid-Wales for six hundred pounds' worth of wool. The mill was to pay my dealer on delivery, and he reckoned that it would take him two weeks to collect, weigh and ship the wool by goods train to its destination. Thus the fortnight delay seemed reasonable, and as the price was good I decided to take the risk. And if one sells direct to Bradford the money is never paid for a fortnight, and the buyers insist, too, on Bradford weighing. There is no imputation against the honesty of the Yorkshiremen, who, like most hard buyers, are scrupulously fair, but I do not like the principle of the thing. I always insist on being present at the weighing.

My wool went off in eighteen great sheets on two lorries, and I followed in my car to a nearby railway station. I weighed it with my dealer, and received my post-dated cheque for over a hundred and fifty pounds. Two weeks later the cheque was returned dishonoured. I went to see the dealer and found him very worried. He had himself received his six-hundred-pound cheque from the buyer, and had paid it into his bank against the presentation of the several post-dated cheques which he himself had given. But the main cheque was dishonoured. The buyer had immediately gone into liquidation. My dealer on behalf of his clients took legal action, but could neither recover the wool nor obtain a dividend on his debt. The income tax authorities had a prior claim on the assets of the mill, and not a penny was forthcoming to other creditors. The mill owner's wife was left in a prosperous condition, as most of the assets were held in her name.

Bankruptcy is a sure road to success if handled properly. It is as profitable as a serious fire. Dyffryn was able to stand the loss of the wool crop, though the small margin between profit and loss in British farming rendered the setback serious. But there were several small farms in my neighbourhood which were staggered indeed by the mishap. A farm today must have a large turnover to be economic, and the smallholding cannot properly carry the occupier, his wife and his children. They receive often less than a labourer's wage for their work and worry, but they are not to be pitied, because they are rich in independence. I continued to sell to the same dealer in after years. He had not been dishonest, only careless. And he has paid me a few pounds each year off the old debt, because many of the smaller farms followed my lead in selling again to him. Our only hope of saving anything was to keep this dealer on his feet. But now I insist on receiving cash in notes for my wool, and I drive home after the weighing to show Esmé sometimes two hundred pounds in cash. Esmé's mind thinks in shillings rather than pounds, and the sight of so much ready money means little to her, but upsets her in some vague way.

However, if I had not been caught this once, I should never have learned the lesson of caution. Experience is the only lasting treasure. I am not a fool, and could possibly have made more of a monetary success in some other occupation. But then most occupations are standardised, and one knows their character by hearsay already, so little fresh can be learned. And, anyway, I had had a bellyful of business methods in my Canadian factory. One business is the same as another. Businesses are like gassy soft drinks, which have the same mentally flatulent effect in spite of the different colourings which deceive one into imagining that the flavours are varied. Experience can never go bankrupt. Many magnates are bankrupt and do not know it. Many politicians live in a mental Carey Street. It is surprising how many statesmen are countrymen. Lloyd George, with his pigs and his fruit, may be a poseur. Cincinnatus was probably genuine. England's first dictator was a Huntingdon squire, and the world's earliest democracy was founded upon the fields of ancient Greece. Country folk are versatile.

My experience with the wool saved me a later loss. Two years ago a man came up to Dyffryn house to see me. We had recently sheared, and the white fleeces were piled high in a building to await a bid. The man had glimpsed them in driving past, and wished to make an offer. I walked down with him, and he handled the wool. He picked up two or three fleeces which had escaped washing, but did not remark on the grease in them. I use pitch to stamp my 'R' on the sheep, and strange buyers complain of this because Bradford has difficulty in removing the substance. But when I tried for one year a proprietary soluble marking fluid I received no better price, paid more for the fluid than for the pitch, and found that the weather speedily obliterated the mark. So I returned to the old method, although a buyer would always try to depreciate the price because of it. But my stranger said nothing. He was unable to estimate the weight of the wool, and I had to tell him that there would be about three thousand five hundred pounds weight.

The man was a queer-looking fellow. He was very tall and burly. I am big myself and not weak, but I felt that the stranger was a difficult man. He was one of those animal types whose aura is purely physical. They always induce a physical reaction in me, and I kept weighing him up as if we were two prize-fighters. He had a bland red face like a Shropshire man, but I could not place his accent. He left the wool pile and swaggered back to me, hands on hips. I felt that his examination had been play-acting, and that he was no wiser.

'I'll give you ten and a half,' he said.

Now there had recently been held a wool auction in the next county. Buyers had taken their pick of the wool at ninepence-half-penny. I knew then that there was something very queer indeed about the man, but thought it just possible that he was a fool or that he wished to establish himself as a buyer in my part of the world. Dyffryn has one of the biggest wool crops in Wales, and no doubt his integrity would be established in the eyes of smaller farmers if he could tell them that he had had my wool. I held out for eleven-pence, but could not move him.

'Well, then,' I agreed at last, 'I'll sell at ten and a half, but you must provide sheets, and must pack and cart to the station for weighing.'

He said that these conditions would suit him. They certainly suited me, because the labour and cartage were worth the half-penny per pound over which we had been arguing.

'What about draft?' I asked.

Bradford always insists on docking about one pound per cent from the weight to allow for loss by evaporation of moisture in transit. This is an absurd excuse, for the wool industry was origi-nated in Yorkshire partly because of the humid conditions which render the staple workable. Wool is more likely to gain weight than to lose it by the time it reaches Yorkshire warehouses. The man did not seem to know what draft was. I explained, sorry that I had mentioned it.

'Oh, we won't bother our heads with that!' he said.

There was one more question. I asked him to let me have a banker's reference.

'Don't worry about your money, old man,' he grinned. 'I always pay in cash.'

Cash is cash. And I could not conceive that his notes would be forged. Besides, this procedure is not unusual. Many people do not use cheques and pass-books because they enable income-tax people to check accounts. I said that this would suit me admirably, and inquired his name and address. He said that his name was Kidd, and that he could be found at the Berwyn Hotel at Welshpool. Welshpool is an important market town on the borders of north Wales and mid-Wales. Mr Kidd promised to send his men to pack the wool next day, and then drove off in a powerful Lagonda saloon car.

Of course, Esmé was very excited over all this. A penny a pound above market price meant an extra fourteen pounds or so, to say nothing of free packing and carriage. But we were both worried. I phoned the hotel at Welshpool, but the line was bad, and I had little satisfaction but of a man who was listening to me and holding a conversation with someone else in the same room as himself. Hasty local inquiries brought no result. Kidd arrived early next morning in the same car accompanied by three of the most sinister blackguards I have ever seen together. One was a tall, lean fellow with a sallow face and an ingratiating air. One was tubby and fresh-looking, like an unfrocked priest. The other was small, hard and twisted. He took drugs. I identified Mr Kidd more and more with his piratical namesake, and set John Davies and Thomas to keep a casual eye on the packing. I do not know if I feared that infernal forces would be called in to spirit away my wool, but felt that elementary precautions were necessary.

The chief pirate came up to the house at lunchtime. We had to ask him to eat with us, and I gave him a drink. I disliked the way in which he looked at Esmé's trim little person. He noticed my coolness, and began at once to entertain us with anecdotes of the Black-and-Tans, with whom he said he had served during the Trouble in Ireland. I disliked the man so heartily by this time that I forgot my duty as host, and tried several times to trip him up over his more spectacular stories. I could not do so. I believe they were true. After lunch I got him out of the house and found that the last wool sheet was being filled. A wool sheet is like a giant's pillowcase. It is about six feet by four, and open along one long edge. The men had slung it from a beam by a rope fastened to its two corners. The twisted man was standing in it to tread down the fleeces as they were tossed to him by the lean fellow. The rubicund cleric was sewing up the openings of the filled sheets with twine and a sail needle. There were seventeen sheets. John Davies whispered to me that the men were very curious and kept asking questions about me.

Kidd regarded the sheets judicially.

'We won't quite pack those on the big lorry,' he mused. 'It'll be a pity if we have to come back for the few left over.'

'Rope a few on top of the car, boss,' suggested the lean man. 'That'll leave just a full load for the truck.'

'We'll try to put three on the roof of the Lagonda,' said Kidd to me, 'and you can follow me down to Bettws station to weigh them. Then the lorry can collect the rest to-morrow, and I'll meet you again at the goods yard to weigh.'

'That'll be all right,' I said. 'You can pay me for these three as soon as we weigh them.'

He did not seem to like this suggestion for a moment, but agreed almost at once. It took a long time and much labour to perch the three great bolsters on the car and to rope them. Kidd was sweating when we had done, and said he would drive with me in my car so that we could stop on the way for a drink while his crew drove the topheavy load down to Bettws. I did not much like losing sight of the wool, but reflected that if they had the wool, I had Kidd. We stopped many times on the way down, and it was not until the bars closed at three o'clock that Kidd would consent to move on to the station. Fortunately he would allow me to pay for nothing, and for once in my life I did not protest very much. I expect I had ten shillings' worth of whisky out of him.

At the station the wool was unloaded ready for weighing. Kidd had become expansive. He insisted on having a half-crown sweep-stake on the weight of each of the three sheets. I knew to within a few pounds what a sheet packed to a certain consistency would weigh. Kidd had no idea, and I won seven and sixpence. While his myrmidons struggled to repack the sheets we totted up the weights, and found that he owed me over thirty-five pounds. He reached for his notecase and offered me nine pounds in ten-shilling notes. I asked him what that was for. He said it was a deposit so that I should know he was acting in good faith and would return for the rest of the sheets tomorrow. I said that in that case it was quite in order, and that I would give him a receipt at once, but that his men had better unload the wool again so that it could wait at the station until the balance of the thirty-five pounds was forthcoming. I was very angry. Most people have a conceit which infuriates them when they are taken for fools. I thought for a moment that the bunch were going to set on me, and I did not much care. But their caution prevailed, for had they got away with the wool they would have been stopped on the road.

Kidd suddenly gave in and ordered the gang to unload once more. I was glad to see that the men hated the hard work. Kidd became friendly and reproached me for lack of trust. Presently they all drove off with a promise to return next day. I saw the sta-tionmaster and obtained a receipt for the wool in my name. He seemed surprised, for Kidd had slipped round to say that he had bought it and was leaving it for the night. But the station-master knew me and took my word. I drove home. Esmé and I were not really disappointed. The whole affair had been an obvious swindle, and we had played with our eyes open. A few minutes after I had arrived home the station-master telephoned. He had just received a phone message to ship the wool by express goods to Welshpool. Was that all right? I told him that it was not all right, and that he was only to deliver it to me in person.

However, I had gained nearly a bottle of whisky, seven-and-six in bets, and seventeen wool sheets.

Kidd worked his trick all over north Wales. I told the police, but they were powerless to do more than warn farmers whom they discovered were in contact with the man. Kidd always told the same story. His lorry was to call next day for the bulk of the wool, and meanwhile he might as well take a few sheets on his car. And he was always careful to pay his deposit. This left the farmer with no redress but to sue in the civil courts for the balance due on the vanished sheets. Such action would have been to throw good money after bad. Kidd was a man of straw, and it is certain that he would have had no tangible assets.

But he left me the gainer not only of whisky, but also of experience.

15

The Hay Harvest

OUR HAY MATURES LATE AT DYFFRYN. It is not ready to cut until after shearing. No stock has grazed since lambing along the flat meadows which lie along the floor of the valley between the road and the river, and it is here that we take our crop. In the old days much more land was mown on Dyffryn. Labour was cheap then, and the art of scything was common, so that rough places could be cut. Most of the ffridd used to be taken for hay, and a man would cut one swath up to the mountain wall, where he would lunch, and one swath down to the house for tea. The thin hay must have been carried in roped bundles on the back to be stacked in the many little round enclosures whose crumbled walls still remind us of the past.

The pygmy efforts of man to fight nature with primitive tools are one of the glories of our species. I have seen the same type of harvesting in progress high on the hanging flanks of the snow-capped Pyrenees. The same indomitable ant-like persistence has forced nature to grant a living to the irritating biped. For as quickly as she dashes him to the ground he scrambles laboriously to his feet again and fights on blindly. In those distant days at Dyffryn, cattle were nearly as numerous as sheep, and the problem of winter fodder must indeed have been difficult. There were many more farms in the valley then, and sheep kept on the reduced area of each farm would not in themselves have produced a living, so the mountain dwellers slaved to store food for the cattle who would give them milk and butter.

The old Welsh cattle were a picturesque breed. There are still a few to be seen, probably throwbacks. They are black, with a white band of creamy white running like a girdle around their body. They are known as belted cattle.

But the rise in the standard of living has produced an inverse alteration in the standard of labour. Long before I went to Dyffryn the hay harvest had become an easier event. A Dutch barn had been built at each of the three sets of buildings, and hay was cut only along the accessible meadows. At my valuation I took over the two-horse mower, and for several seasons we used this implement over most of the hay area. The rougher parts are cut by John Davies and Thomas with scythes. Hay time is heartbreaking in the hills because of the unreliable weather. We dare not cut much grass at a time in case rain comes and persists long enough to ruin a large area of cut hay. Sometimes it rains just as a big piece is ready for carting. Next day, if it is fine, we spread the rows to dry again in the morning. But as soon as the hay is once more gathered together in late afternoon ready for the carts, the rain sweeps down the valley again and the routine has to be repeated next day. And perhaps each day for many days, until the blackened, rotting grass is fit only to be raked in heaps and burned.

I used at first to engage two or three extra hands for hay harvest. But it became increasingly difficult to obtain good farm labour, though harvest wages before this second German war were up to fifty shillings weekly, plus full board and lodging. And it was unfortunately true that many of the men on the books of our labour exchanges were unemployable. The fault may not have lain so much with the men as with the economic system of the country, but the result remained. And John Davies and Thomas did not work smoothly with these birds of passage. They despised their shiftless ways and their artful expedients for dodging labour. Afterwards the great rearmament programme drew more and more farm men to the building of aerodromes and the erection of camps, and later, of course, conscription further complicated matters. The Army were not disposed at first to release men for agricultural duties. Farm workers might fight Hitlerism, but they might not undertake the more skilled work of staving off that even more implacable foe, starvation. Now, however, wisdom has entered into our counsels.

After a few seasons I decided to improve the bottom lands. The Government subsidy had come in force for lime and for basic slag. On lime the Government paid half of the delivered price, and on slag one quarter. Our peaty upland soil was very acid, and the lime sweetened it so that the finer grasses, further aided by the slag, began to oust the coarser. But it is useless to fertilise damp ground, and these flat lowlands held the water. Much draining was necessary, and even after ditches were cut the spongy sphagnum moss which grew among the stiff tussocks of fibrous molinia grass kept the ground too moist. Sphagnum will hold twelve times its own weight of water. So one winter I decided to buy a very expensive harrow with which to tear out the moss, break up the tussocks, and aerate the ground. There is at fifteen miles from Dyffryn an agricultural college whose advisory staff are second to none in Britain. They had advised me about my fertilisers, and now recommended a special harrow for my purpose. This harrow was a two-wheeled implement eight feet wide, and its two rows of spikes were dropped into and lifted from the ground alternately, worked on the cam-and-roller principle. As one row of spikes scored through the mat of vegetation, tearing out moss and decaying grass, the other lot of spikes was automatically cleaned of its load of debris. This harrow was excellent, but on our rough land a tractor was necessary to pull it.

We hunted everywhere for a second-hand tractor in good order. One was not to be found at a reasonable price. But one day Esmé and I were in mid-Wales to buy rams. I glanced at a local paper and noticed that a slate quarry had a Fordson tractor for sale. We drove to the quarry, and I was led down a subterranean passage to view the tractor. It seemed in very good condition, to outward appearance at any rate, and I was told that it had been used almost entirely for pulley work. I knew nothing about tractors, but kept silent and looked knowing, so that the quarry manager himself anxiously supplied information item by item as my expression grew more and more disapproving. The quarry was being fitted with electric power, so I knew they had no further use for the tractor, and that they would have great difficulty in that rugged district in finding a customer. I abruptly offered fifteen pounds. This was an insult, but the manager turned the other cheek by accepting with such alacrity that I felt I ought to have offered only ten.

John Davies regarded the newcomer as St George must have eyed his dragon. Thomas sprained his wrist by swinging at the starting-handle with the ignition advanced. But the tractor was a success, and the harrow too. 'Indeed the flamer have got some guts!' exclaimed John Davies, as he saw the great rear wheels ploughing relentlessly through marsh, scrub and hillocks of grass, with the harrow operating behind. Our manuring and harrowing improved the existing hay-lands and opened up more. And as the prospect of a heavier crop grew, so the labour difficulty became more serious. I decided to mechanise. Our neighbours, who had long suspected our madness, were confirmed in their suspicions.

We adapted our old horse-mower so that we could attach it to the tow-plate of the tractor, and John Davies, who loved the old mower, sat gingerly on the precarious spring seat to watch the blade among the rocks, ditches and tussocks.

If John Davies did not quit my service then, he never will. Day in, day out, he careered willy-nilly behind me, clinging somehow to his leaping iron steed. Sometimes the tractor would negotiate a ditch which proved too deep for the smaller wheels of the mower, and a startled exclamation would cause me to look round and then hastily to pull up. The short shaft of the mower would be pointing despairingly at the sky, and Davies, tumbled backward off his eminence, would be fussing about his charge like a mother reassuring her frightened child. And every now and then parts of the mower, which was used to a much more sedate pace, dropped off, and we would spend frantic minutes hunting in the cut grass for stray rods or nuts and bolts. The continuous violent jolting and the fumes of paraffin from the tractor's exhaust had a bad effect on John Davies's health, but he never complained, except once when I saw him look particularly green in the face. He only said, 'Indeed, I do have a cramp in my belly.' Thomas spent much of his time scything the more inaccessible places, and Esmé had Belle, the mare, with the swath-turner, to spread the new-mown hay. And when it had dried in the sun or the wind she would transfer Belle to the hay-rake and gather the grass into long windrows. Then I would unhitch Davies and his Frankenstein and fasten the hay-sweep to the front axle of my tractor. The hay-sweep which I had bought was much too light for tractor work and for our rough land. It was really intended to be used on a car, so I was constantly breaking the long tines as they slid along the ground in front of me. They would stick into hillocks, break on stones, or smash as I turned with a weight of hay on them. But even so we swept our hay to the Dutch barn, where Thomas and John Davies stowed it safely away. At Dyffryn one always feels that each load is a load snatched from under the nose of the malignant weather.

After our first season with the tractor we became confirmed mechanics. At odd times during the next winter we disposed of the many ditches which were such obstacles to tractor mowing. Many of these had been cut haphazard in the days of scythes, when they had not been obstacles. We dug them out, then filled them with stones so that they would continue to drain. When turf was stamped down on top of the stones the surface of the ground gave an unbroken sweep for our tractor. And I decided not to impose further on John Davies's stomach. I bought a tractor mower. I did not get the trailer type, because in our rough country the operator needs the blade under his eye so that he can ease it by the lifting lever over rocks and inequalities of ground. I bought instead a bar which was attached to the tractor in front of the off rear wheel, and whose cutter was driven from the power take-off. I sold my inadequate sweep and bought a tractor sweep. This was twelve feet wide, and the tines were ten feet long. A lever came beside the tractor seat, and I could raise the points of the tines from the ground as soon as the sweep had picked up a load, so that it was easy to manoeuvre into the yards. But for Thomas I found the best toy of all. This was an Atco Autoscythe. The heart of the machine was a Villiers two-stroke engine which would run for two hours on a quart of petrol. The engine drove two parts at once. In front of the implement was a rubber-tyred wheel which dragged it along, and projecting to one side was the three-foot cutter bar. The willing little motor would pull the machine, cutting as it went, up a forty-five-degree slope. Thomas had simply to walk behind to guide the course by two grips like plough-handles. He and his father took it out as soon as it arrived to try it on bracken. They were by now confirmed modernists. Both came back with their eyes shining.

'It do work a treat, for sure,' said Thomas. And it did. While I raced up and down the meadows on my tractor, making a six-foot cut at ten miles an hour, Thomas chugged steadily over the small, rough patches to cut as much as five acres in a day. John Davies worked Belle in the swath-turner or the rake, and Esmé presided over all of us with a benign smile.

'I do feel like a gentleman,' announced John Davies. 'There ben't no work for a fellow.'

The sweep was another delight. I would ravel up fifty yards of windrow, until the heaving hay mounted as high as the radiator cap, and would whisk it to the barns as fast as the men could handle it. John Davies had to become a labourer again, and he and Thomas, under their many layers of clothing, must have sweated off pounds of weight when they were stacking. As for me, I rode my clanking charger in only shorts and boots, cheered by the occupants of the summer charabancs. There is nothing so beautiful as a brown, fit body. Or am I an exhibitionist?

I had usually to begin my mowing alone at about six in the morning. One day I did my two-hour spell and then lifted the cutter bar to cruise back to the building, where I wished to leave the tractor for Thomas to grease while I went for my breakfast. The cutter bar weighs over a hundredweight, and is also kept to the ground by a powerful coil spring. The long lifting lever, when the bar is up, is drawn back so that it is just clear of the operator's right side. That morning I must have handled the lever carelessly, for the pawl did not properly engage in the ratchet. At any rate, as I leaned sideways to see if the bar was well lifted, the lever sprang free and struck me behind the ear. I tumbled unconscious out of the seat. But my mind must have told me that my collapse was premature, for presently I sat up and saw that the tractor was still proceeding towards the building. I got to my feet, raced after it and switched off the engine. Whereupon I lay down once more to complete my interrupted faint. Thomas, who saw the incident, said that it looked a treat, and recounts it still with much amusement.

But his own turn came. It is only remarkable that he had not mangled himself sooner. He was strolling along one day behind his Atco, when he stopped to make an adjustment. There is a clutch on the Atco which disconnects the drive to the wheel, but which leaves the cutter bar oscillating. Thomas bent down to put his hand to the machine. The Atco must have been on a side-slope, for as he stooped the whole thing tilted. The bar swung upward and caught his finger in the moving knife. I knew nothing of this, although I noticed that Thomas had a dirty rag wrapped round his hand all day. But as he usually has some fresh wound I took no notice. Towards evening he came to ask if he might stop work to go to Bettws. He and his father are so excellent and conscientious that I always think my best return is to let them feel that they can take time off for the asking. I said that Thomas could go, although I did think it a little curious that he should desert us in the middle of the stacking. As he left the field his father jerked his head after him.

'He do potch too much with that Atco,' he said crossly. 'I have tell him.'

I did not quite see the relevance of this remark, and said so. 'The silly bugger have slice off the top of his finger,' explained John Davies, tossing up another pile of hay.

And he had, too.

Now that we had taken the measure of our hay harvest and were defeating nature with science, Esmé had an Idea. She had read somewhere of a new system of haymaking adapted from North Country and Scandinavian methods. A company had been formed to exploit 'Pyramid Haymakers'. The principle was that one erected these Pyramid Haymakers at convenient intervals and stacked around each about half a ton of hay. The tripod erection left the centre of the little domed stack hollow, and air entered this space through an opening raked away at the base of the pile. The circulation of the air matured the hay, and the shape of the stack, which was like a large haycock, rendered it impervious to weather. The system certainly seemed to have advantages in our mountain climate, for it was claimed that the hay could be stacked almost as soon as cut, so that the interval during which the weather could spoil it was cut to a minimum. Even with our light and wiry upland grass it would usually be thirty-six hours or so before we could sweep in cut hay, and in that period of time we might experience anything from hail to a cloudburst. But with the Pyramid Hay-maker in use the interval was cut to a few hours, and once stacked the hay was saved, whatever the weather might do. And after a month or so one drove the sweep under the pile and carried it to the barn.

It appeared that Esmé had written for pamphlets. But in their place appeared two salesmen. They arrived late one autumn after-noon. One was a retired Regular Army officer, who introduced himself as Colonel Cante; the other was a hard-bitten colonial, who was obviously a rolling stone and a tough nut. His name was Wace. We asked the visitors to join us at tea, but found it difficult to extract any information. The colonel was an ardent member of the Oxford Group, and Wace was a Jew-baiter. The two men disliked each other heartily.

The colonel embarked at once upon his life history. He was apparently long married to a charming wife, but had for many years deceived her about his infidelities. She had known nothing of his numerous affairs and had been very happy. However, the hand of Providence had led her husband to Buchman, and the colonel began to indulge in an orgy of introspection. Finally he cornered his wife and came clean to her. To his surprise she was upset by his revelations, and to allow her to become attuned to the new philosophy he had left home temporarily. At intervals he kept remembering further indiscretions which he had not mentioned. He wrote home about these, giving circumstantial accounts. His wife's attitude became still more unreasonable, but the colonel had no doubt that time was on his side, and that she too would soon evince a desire to come clean. He was looking forward to plumbing the depths in a spiritual tête-a-tête with her.

Wace was less intelligible. He was muddled between the San-hedrin and the Protocols, and had frequently to check the accuracy of his monologue by a thick volume which he carried entitled *The World Menace*.

At intervals Esmé or I would inquire the price of the Pyramid Haymakers. The colonel would say that they could be sold more cheaply if we were all to come clean and to cohabit in love. Wace would say they were too dear because of international finance, whose system had been planned by King Solomon in the Temple. The salesmen showed no signs of leaving. We took them down to the village for a drink, and the colonel undertook the conversion of two Calvinistic Methodists who attempted to wash away their sins in the beer which he provided for them. Wace distributed several tracts, and drank himself into a brooding fury. Esmé and I, for once both at a loss, carted our men back home again and gave them supper.

The colonel showed no signs of leaving us, but sat by the fire and radiated brotherhood, smiling gently. Wace had by now stimu-lated himself into a fighting rage, and kept producing leaflets from all over his person with the adept air of a conjurer. We asked the two salesmen to stay the night. They accepted in full stride, scarcely interrupting their two incongruous streams of discourse.

I left early the next morning for an auction which I had not previously intended to visit. I was sorry to leave Esmé, but I felt that she had brought this infliction upon herself. When I returned cautiously at dusk, the prophets were gone. But Esmé was very quiet. When I asked questions she prevaricated. Stacks of booklets on love or on hate were distributed about the house. At bedtime the secret came out. She had bought eighty Pyramid Haymakers. I was suitably angry, but secretly considered the expenditure of forty-five pounds justified, for at any rate we were left to our Evil ways, and to our indifference to the growing Menace. But Esmé's caution had withstood the stress of the visit, and she had extracted a promise that the Pyramids could be returned at our expense if unsuitable, and that if we decided to keep them we could pay for them over three years.

The colonel and his companion had left, the one smiling, the other scowling, like cathedral gargoyles.

John Davies did his best next season to make the Pyramids a success, and Esmé fluttered anxiously around her brain-children. But we found that we could not stack our hay upon them as quickly as we had been told. If the grass was stacked too green it became as mouldy as it would have become in a barn, so we had to leave it a considerable time to wilt. We might as well have left the hay a few hours longer and have swept it to the barn, where it would have been finished with. And the legs of the Pyramids sank into our peaty ground under the weight of the hay. Once a gale of wind scattered the grass from them until their bare bones were exposed. We used thirteen, and on that mystic number gave up. We returned to the bad old ways, and Esmé was for many days distrait.

At last she said, 'When you return those Haymakers you bought, let me write the covering letter.'

16

The Snack Bar

LATELY OUR SUMMER HAS BEEN made even more busy by the addition of another responsibility. Esmé has little time now for her games with Belle and the hay rake, for August is the very month when she is kept most busy tending the child of my Idea. But it was from Esmé herself that I caught the Idea infection, so she must not grumble. Three winters ago, early in the New Year, Esmé and I were walking high up in the ffridd with Luck and Mot. It was a frosty clear day with a pale sun, whose beams were reflected in the light fall of snow. The bracing northern climate has made the Nordic races pre-eminent. And the nip in the air made me think on that day. Suddenly I said, 'I have an Idea.'

Far below lay the twin lakes, for once unruffled by wind, and Siabod was faithfully duplicated in their icy blue mirror. The river tumbled into the lakes, frothing down a waterfall and some rapids before it split to surround an island which would later be mottled blue and gold with hyacinth and broom. A quaint little flat bridge spanned the bend of the fall, and its drystone piers were pointed to cut the torrent. Opposite the bridge the road ran flush with a small level piece of ground, perhaps an acre in extent. For the rest of its journey through the valley the highway was hemmed by stone walls or by steeply sloping ground, so that it hurried on apologetically. But here it had a moment's rest in the spaciousness. I drew Esmé's attention to the fall and the bridge and the patch of ground. She is always rather on her guard against Ideas which do not bear the authentic stamp of herself as the original manufacturer, and she looked suspicious and non-committal. I said, 'You know how visitors in the summer always park their cars on that patch to picnic. It's the only place for miles where they can draw off the road, and it's a lovely little spot. It's nothing to see six or eight cars there on a nice summer day. Let's build a roadhouse.'

Esmé was shocked to the core of her being by the magnitude of the conception, but tried not to show it.

'How much will it cost?' she asked.

'About a thousand pounds,' I answered, not knowing at all. 'It's a lot of money,' said Esmé primly.

'Oh, we'll borrow it!'

'Who from?'

'Plenty of people would think it a good investment.'

'Oh!' Then: 'What shall we call it?'

More Ideas welled up.

'We'll call it the Lumberjack, and we'll build it from logs, with a cedar-shingle roof.'

'We'll build it?'

'I will – with John Davies and Thomas.'

'Do you think you can?'

'Oh, yes. We were always building log cabins when I was in Canada.'

'But a thousand pounds!'

'And people will come for miles around for lunches and teas; the situation is unique. The lakes, the waterfall, the old bridge, Snowdon. The AA told me the other day that two thousand cars a day pass through the valley in August.'

We scrambled off down the ffridd and were presently inspecting the site minutely. We laid out a bowling-green, a putting-course, a bathing-pool, and a hard tennis court. Esmé pointed out that we could keep people in the valley all day if we gave them enough to do.

'We can advertise four square miles of shooting,' I agreed. 'There are those two hares under Glyder Fach, and Matthew, Mark, Luke and John, the four grouse on top of the ffridd.'

'And four miles of fishing.'

'And we'll build a terrace right out over the falls for people to feed on.'

We went back home greatly excited, and I drew many plans.

But presently we both fell quiet. At last Esmé said, 'What about our lovely valley?'

I had begun to think of that myself. The fine remoteness of the place was not lightly to be bartered for pieces of silver. But the aesthetic does not produce calories, bread and butter does. And farming is a bleeding sacrifice on the altar of industry. I tried to salve Esmé's conscience and my own by laying the blame for our necessity at the door of a Government which existed for industrial interests only. But we were both uneasy.

The many plans which I drew became more and more elaborate. At last Esmé said diffidently, 'Let's have an architect down, just to make sure everything's all right. If the roof fell in...'

She had a fey look in her eyes. I agreed about the architect, pretending to be hurt, but secretly very relieved. If the roof fell in... We wrote to friends in London, and they recommended a rising young man named Bertram Pollock. In response to a letter Mr Pollock said that he would be delighted to stay with us for a weekend and to give us his advice. One wild wet night towards the end of January we went to a railway junction twenty-five miles away to meet our guest. Esmé and I both have a liking for fast cars. We run a fast car until we are beggared, and then revert for a while to very second-hand baby Austins while our finances are recouped. We had at this time our third Bentley. It was an open car, and we dared not put up the hood in case the squally wind should rip it off. And besides, we had long since found that if we drove at a sufficient speed the rain cleared us and left us dry. Mr Pollock had arrived at the station, and was standing just within the waiting-room. He was regarding the howling, splashing night as Shadrach and his companions must have eyed the furnace. He was a tall, thin young man of about my own age. He was pale, with a long, pointed nose and longish hair. Not much else was visible as we shook hands with him, for he was muffled in a greatcoat and scarf, and his hat was pulled over his eyes. He peered uncertainly at us as if through the slit in a visor. At times I have an affectation of driving in winter weather wearing an open shirt and shorts. The principle of this madness is on a par with the Swedish expedient of leaping out of a steaming public bath to roll naked in the snow. I was nearly naked that night, and our visitor's eyes came out on stalks to stare at me. We explained to him about the danger of putting up the hood, and assured him that we would try to travel sufficiently fast to keep him dry. We put his suitcase in the back of the car, and coaxed him into the narrow front seat with us.

In a hotel near us there was a farewell party being given that night to some climbers. I asked Mr Pollock whether he would mind joining us at the party for a little. His reply was inaudible above the roar of the engine and the screech of the storm, but I took it that he was agreeable. One of his gloved hands was clamped on top of his hat. The other arm was curled behind Esmé to clutch the back of my seat. Both feet were pressed rigid against the floor-boards. Our treatment of our guest was indefensible, but it was due to thoughtlessness rather than malice. We were well used to our rough life. I was strong and fit, and Esmé, for all her air of a Dresden Shepherdess, is physically as hard as nails. I knew the road home so well that I could almost have followed it blindfold, but I did not realise that to a stranger our tearing progress must have seemed suicidal. The road lies through rocks and woods, and is so twisting that the headlights shine full on rocks or tree trunks or over little cliffs, rarely lighting more than a few yards of tarmac at a time. And we were in a hurry to reach the party. The car could exceed ninety miles an hour, and we used most of our ninety on several of the brief straights. Poor Pollock, fresh from his comfortable London flat, must have felt that he was being whisked through hell by a half-naked demon and his grinning familiar. For Esmé always smiles when she is doing something rather exciting.

We reached the hotel quickly, and propelled Pollock's rigid body into a crowded room of muscular men, to whom servitors were carrying beer in milk-pails. His stiffened fingers remained clenched, and his glass chattered against his teeth. He crouched in a corner with apprehensive eyes, as if fearing an attack. And it was not really until after he had gone back to London that we began to realise how very alien our world must have seemed to him. In time we dragged him out into the gale again and drove him home. He gasped audibly as we swung unexpectedly off the road and roared slithering up the mountainside to Dyffryn house, with our bright lights pointed skyward to cut grey channels through the racing squalls of rain and cloud. And next day was a real Dyffryn day. The wind screamed about the walls. The solid house quivered dully. The rain flung itself in sheets against the windows. We stayed in till afternoon, but the storm only grew more violent as it prowled about in wait for us. Pollock had to return by an early train next morning, so there was nothing left for us to do but to dress him up in some of my oil-skins and to lead him down to the proposed site. In the wind his thin body bent like a reed, and his hat sailed spinning down the valley. We showed him where we wished to build, and we held him up as he tried to survey the place. He was conscientious and did his best, but he was physically unable to move about, and his quiet voice was lost in the plunge of the flooded fall and the screech of the gale. I did lead him near the fall, and I believe that for a moment he thought I was going to hurl him down in some quaint survival of Druid ritual sacrifice. His body stiffened hopelessly until I yelled 'veranda' and 'sun parlour' at him. But 'sun parlour' startled him the more, and he threw a wild glance at the furious elements. We got him to the house again and sent him for a hot bath before tea.

Pollock left by an early train next day. We were rather late starting for the station, and I had to hurry. It was, of course, daylight, but I think he preferred the night terrors, at which he could only guess, to the terrors of the day, which he could see. I last saw him snuggled into the corner seat of his train, his eyes happy with the incredulous light of a reprieved prisoner. In London he spent a few days locked in his rooms, as we heard, in a state of collapse, then he sent us his plans. His work was very sound and comprehensive. The place we had had in mind would have cost fifteen hundred pounds equipped. But when Esmé and I worked out the overhead expenses we were afraid. It would only be worth while to be open for the few summer months, and the interest on capital and the maintenance costs for the other eight months of the year would alone call for a large turnover. And then there was the difficulty of persuading staff to come to such a remote place. We decided that our grandiose scheme was possessed of too much risk.

'I'm going to have a sectional wooden hut built,' I said, 'and we can put it up each summer.'

Thus our valley would be clear during the real months of the year. For when the madding crowd swarms through in summertime to prostitute the hills, a building more or less is of little account. But when we are left alone to our country we are very jealous of her appearance. I ordered a building eighteen feet long and eleven feet wide, made of weatherboard and with a plain ridge roof. A three-foot veranda ran the length of the front under the roof, and so eight feet were left for counter and serving space. I paid twenty-eight pounds for the building, but it was not delivered until after Whitsun. We set it up on a foundation of heavy railway sleepers, to which we bolted it. Then we painted the outside an innocuous green, and the inside cream. We did not quite know what to sell. We went to a local wholesaler and bought boxes of chocolate bars and blocks. It seemed that our facilities would not run to weighing, so that everything we chose was ready wrapped. Most twopenny sweetmeats are put up in boxes of either two or three dozen, and Esmé and I still had the private mentality which was appalled at the thought of so many sickly confections.

'What shall we do if we can't sell them?' asked Esmé, her frugal soul upset.

'Eat them!' I answered. And we both nearly retched.

In addition we bought some cases of mineral waters, several hundred cigarettes, some matches and some packets of biscuits. The day came in mid-June when we took down the long shutters and exposed our wares to the world. The purchases which had so daunted us seemed pitifully small now in the vast desert of the eighteen-foot counter. The lad whom we had engaged to mind the place looked tiny too. Esmé popped inside and served the first customer. He was a stout, sweating youth who had ridden his bicycle from Liverpool, eighty miles away, that morning. He asked for a bottle of cream soda, and between gulps inquired which was Snowdon. He did no more than glance at the mountain's virginal severe beauty before mounting his iron steed to pedal back furiously whence he had come. His pilgrimage had been fulfilled. And neither did he realise how momentous was his purchase. His was the first money we had ever taken in trade, and it was the first time that we had tasted the satisfactory joy of retailing. But it was difficult not to think of the four coppers lying in the till as net profit.

June wore into July. We increased our variety of chocolate, and added tins and bottles of all sorts of sweets. Esmé buckled to, and tentatively offered homemade sausage rolls, Cornish pasties, sponge cake and scones. But it was soon as much as she could do to keep up with the demand. We added a tea-urn, which we heated over a Primus stove, and soon found ourselves rushed out on cold days. The profit on chocolate and cigarettes as bought through a wholesaler was something over 25 per cent. Minerals were much better and showed nearly 100 per cent. But tea and the homemade food were better still, and I stoutly maintain that the customers had better value for their money than they would have obtained in most places. People seemed glad to avoid the dreary formality of an eighteenpenny hotel tea with its accompanying delay and restrictions. They praised our quick, alfresco service, and found that they could have a more satisfying meal at less money than they were used to paying. August Bank Holiday came. Esmé had mass-produced unending rows of her snacks, which she aligned in military fashion along the slate slabs of the dairy. These were taken down in baskets, and their ranks melted away before the withering attack of the trippers. Cycles lay on their sides in scores. A dozen cars were always stopped in front of us. Groups of walkers stood to eat our fare. We had to press Thomas into service besides our boy, and Esmé and I served and took money and gave change and answered questions and collected teacups till we could scarcely stand. Our boy, who was paid on a commission basis, was frantic with excitement, and changed his ambition from a watch to a bicycle. We took twenty-seven pounds over the weekend.

August faded into September, and the end of September found our valley left to us alone. We totted up accounts and found that we had made fifty pounds net profit. This allowed for 10 per cent. depreciation on our hut, but not for the capital cost. But we had purchased outright several pounds' worth of fittings. We were very satisfied. At odd times through the winter we laid our plans deeply for next season. We decided to open a week before Whitsun. We knew now how much goods we dare order, and we placed the maximum orders for best discount direct with the manufacturers of every article we sold. I took the car to Bettws station and fetched eight hundredweights of chocolate. And we added postcards to our list, and ice cream. The ice cream was a real triumph. A local factory supplied daily the village of Capel Curig, but it needed the utmost diplomacy to persuade them to give us a freezer and to send on their lorry each day with more ices and freezing material. Then we fitted a compressed-gas apparatus to boil our water, and in retrospect the roaring, vicious Primus of the year before seemed a nightmare.

Esmé insisted on buying a glass showcase for her homemade wares, to which she added several new lines by diligent study of some cookery books. Our counter now was in two sections. One half was given up to proprietary sales, the other to homemade. On the one side we had comprehensive displays of the chocolate of six leading manufacturers. We had a score of other sorts of sweets. And ranged on shelves behind were the main brands of cigarettes. The other half of the wide counter was dominated by Esmé's showcase filled with her fresh products. Around it were stacked various kinds of eatables which we considered more suited to a meal than chocolate. The whole display was most enticing, and was set off by hanging frames of postcards. All children like to play shop. Esmé loved the game, and I surreptitiously abetted her.

With the bar in its improved guise our sales increased from the very start. The Whitsun holiday brought us more money than the August holiday of the year before. Sunday was always the busiest day of the week, and Thomas used to dress up in highest suit to help serve. Each week he was quite overcome by emotion when we presented him with a supply of cigarettes to repay his help.

'I don't want nothing,' he would protest. 'It do be a treat to see the money come in.'

One cannot buy loyalty; one can only reward it.

One day a young man stopped for some tea. He was walking, and his pale physique proclaimed him city-bred. Esmé and I were hanging round the snack bar to watch the receipt of custom, as we often did when not busy. We began to talk with him. He was making a walking tour of north Wales, and was by profession an architect. He came from London. We are always interested to find what brings people to our part of the world, and we asked him why he had come.

'An architect friend of mine', he explained, 'came somewhere round here on an advisory job. He had some extraordinary experiences. The people who employed him were so peculiar that I've always meant to find out more about them.'

'How interesting!' chirruped Esmé excitedly. 'That must have been Bertram Pollock!'

But the young man set down his teacup sharply, and gazed at us with the dawning horror of a wild surmise.

'Oh, my God!' he whispered. And he walked abruptly away up the valley.

The second August Bank Holiday came. The weather was beautiful, and the hills were gracious, remotely condescending in a haze of warmth. Visitors came through in dozens, scores and hundreds. Our snack bar was the centre of a seething mass of raw, sunburned humanity. From early morning till dusk we laboured to take their sixpences and threepences. The weekend realised forty pounds. And in the month of August we took nearly two hundred pounds. When we closed down at the end of September our net profit was a hundred and thirty-two pounds. From time to time we have been abused by strangers for commercializing our valley. These people walk freely over our hills. The country where we make a living is to them a playground. They are made free of more intimate hospitality when wet, lost, or hurt. We suffer damage from them with their unthinking city ways. And if the vote of the masses counts as heavily in estimating the popularity of our little hut as it does in returning a leader to Parliament, then we are public benefactors.

17

The Caravan

ESME WAS A LITTLE UPSET about my Snack Bar Idea. She felt, I think, that Ideas were her prerogative. During the winter which followed the first successful season of the snack bar she showed many signs of abstraction, which presently culminated in a mysterious correspondence with places all over Britain. Letters addressed by most illiterate fists arrived daily for her, and she filed their contents carefully. At last she said, 'I'm going to buy a caravan with my own money.'

I advised her to get one light enough to be towed up the hills of our Welsh roads.

'Oh, I don't mean that sort!' Esmé explained contemptuously. 'I mean a proper showman's caravan. If you can make money out of campers, I can.'

At Dyffryn I have encouraged summer campers by advertising in all the open-air journals. We have increased our takings, at sixpence a head nightly, from a few pounds to thirty pounds a season. It always pleases me to receive each sixpence, because it is such easy money. There is no outlay and no trouble.

'I'm going to let the caravan to campers,' added Esmé defiantly. 'I put an advert in World's Fair, and I've sorted the replies. There are three which sound quite nice near Manchester, two at Birmingham and one at Burnley.'

'But what on earth gave you the Idea?'

'Oh, I just thought of it!'

Esmé is nothing if not feminine. But there was something troubling her. 'You know all about wheels and things, don't you?'

'Well, not all...'

'I'd be glad if you'd come with me to look at what they call the underworks. This one at Birmingham is a fourteen-foot Mollycroft with belly-box and blooms—'

'And what?'

'Blooms. It says so here.'

It did. The letter stated in violet ink that it was a lovely van, with belly-box and blooms. A postscript added that it was also a Mollycroft with Hostess.

'What's this Hostess?' I asked.

'I don't know. Most of the caravans seem to have one.' 'Well, they can't sell her with the van these days, even if it has got blooms.'

But Esmé cannot be sidetracked.

'Will you come?' she asked. 'I'll pay all expenses.'

I had some business to do in Manchester, and readily agreed. We drove one Friday to Manchester, and Esmé took the car and went to see the local caravans while I attended to my business there. She returned so late to the arranged meeting-place that I asked her if she had had much trouble in finding her way.

'Oh, no!' she answered. 'But I had to walk a good deal, because I thought they'd put the price up if I left the Bentley anywhere where they could see it.'

One of the vans had been obviously ramshackle, and the other one, though nice, was priced too high at sixty pounds. The third was travelling with a fair, and had last been seen heading towards Liverpool behind a traction engine.

'And, by the way,' said Esmé, 'the Hostess is a make of stove.' I told her that I had booked a room.

'Where?' she asked, worried. 'At the Midland,' I told her off-handedly. Her thrifty soul writhed, but she gulped bravely, and said nothing but a small 'Oh!'

We dined sumptuously that night and slept in luxury. I left Esmé to settle the bill while I went for the car, and when she joined me she was very quiet. We made towards the East Lancs Road, and presently were speeding towards Liverpool in search of the third van. We scanned each piece of waste ground, and inquired frequently of corner-boys if they had seen a fair pass.

Unfortunately, as always, each man we asked was afflicted with a serious disability. One had a cleft palate, one stuttered, one was stone deaf, and another excelled himself by indicating manually that his teeth were stuck together with toffee. At last a council roadman directed us excitedly up a lane: the panting Bentley gives always an impression of haste, and the man felt he was assisting at a sort of race. And, sure enough, several vans were parked in an acre of mud. Esmé picked her dainty way to a clump of strapping women who were erecting a swing-boat stand. She inquired for Mr Moggridge, and looked at them all so charmingly that the roughest of them put down her hammer and led her personally to a little ten-foot van, where she screamed, 'Joe!' A head appeared over the half-door, thin and sharp-featured.

'That's 'im!' exclaimed the guide, with the air of a wizard, and ploughed away again to her swing-boats. 'Mr Moggridge?' smiled Esmé.

'That's me! What was it?'

'I wrote to you about your caravan—'

'Well, blow me down! Emmy!' Another head shot out. 'This is 'er what wrote.'

Emmy was so billowy that I wondered Mr Moggridge was not afraid of being overlaid.

'Well, strike me pink!' exclaimed Emmy. 'And what does you

want a van for? No offence.'

But Esmé gives away nothing, and intimated that she wished to sleep out for the sake of her health. She looked radiant in health as she stood there, her elfin face aglow, her slim body lithe and hard. But Mr Moggridge thought her intentions admirable. He gave a peculiar penetrating whistle, and five children of indecipherable sex materialised from the mud.

'Look at 'em!' he exhorted us. 'Fourteen years old dahn ter six, and nary a day in bed. Them three big 'uns sleeps with Granny in th'old bus, and the two nippers mucks in 'ere.'

At this Mrs Moggridge appeared again from the interior and coyly invited us to step in for a cup of tea. Mr Moggridge barred our entry for one moment.

'It ain't just any Tom, Dick and Harry as she'll ask inside,' he whispered, nodding significantly.

The van was small and cramped. Lockers, cupboards and a folding table occupied one side. The tops of the lockers were cushioned for seats. The little, but highly efficient Hostess stove occupied the other side, flanked by more lockers. The far end of the van was filled by a double bunk, and in the dark recess underneath it the children evidently 'mucked in'. There was a neatly curtained window in each long wall. We sat to drink our tea, and Esmé caused further astonishment by asking for a glass of cold water. She explained that she had never drunk tea in her life, and did not like its smell.

'That's what gives 'er 'er complexion,' observed Mr Moggridge admiringly. 'We drinks tea mornin', noon an' night, an' look at us!'

After the tea ceremony we were shown the intricacies of the van. While Esmé was diving into cupboards Mr Moggridge took me outside to watch him thump the underworks. He pointed with pride to the capacious belly-box, which was a locker-like contrivance slung between the axles.

'Put all me pots and pans in there when we're on the road,' he assured me.

Then he drew my attention to the four rubber-tyred wheels, and claimed that the caravan would tow so easily that I would not know if it was behind the car. To prove this he heaved at the drawbar in spite of my protestations, and did in fact succeed in moving the van a little, but with such abruptness that Emmy thrust a red face out of the door to demand what he was playing at. At last Esmé came delicately down the steps, assisted tenderly by Mr Moggridge. She said that she liked the van, but that she had promised to see some others before coming to a decision.

'Right and proper,' agreed Mr Moggridge. 'Right and proper. Take your choice. We won't 'old it against you – never.'

And on these sentiments of mutual esteem we squelched back to the car. As we headed for Birmingham we discussed the van. It was fairly cheap at twenty pounds, equipped as it was with pneumatic tyres, but it was small, and the fittings needed a good deal of carpentry.

'Mrs Moggridge said it hadn't any ticks,' Esmé told me. 'Should it have?' I asked.

'No, but she says many vans have, and that if you once get them in they're buggers to get out.'

'You're being too literal,' I said. 'Mrs Moggridge's metaphor was allegorical.'

We arrived in Birmingham, and I drove to the Queen's Hotel. Esmé half uttered a protest, but in the end shut her lips unhappily. As I hesitated at the reception over the question of a private bath she had the air of one who thinks, 'This is the end!' I did not take the bathroom. Next day she paid the bill once more, and we drove towards a disreputable quarter of the city. After intricate navigation and much questioning of people who mostly turned out to be strangers to the town, we came to a waste plot of ground surrounded by a high hoarding.

'This is the place,' whispered Esmé, as if afraid of being over-heard above the thunderous engine. 'Drive on and hide the car.' I drove on and turned into a side street. But as we left the car, Esmé had another qualm.

'Everything will be stolen out of the car if we leave it here,' she said.

So we drove back to the hoarding, and Esmé slipped furtively in front of me through a postern door set in vast padlocked double gates. There were several vans and converted buses inside, but only one seemed to be occupied. As we approached this, a man came out and posed majestically on the front platform. He was about sixty years of age but held himself regally. He was as tall as I and very broad. His hair was dyed black, and flowed down his neck like a mane, and he wore a tremendous Kaiser Wilhelm moustache. He was dressed in cream breeches, patent-leather black top-boots, and an extraordinary-shaped black frock-coat.

'Me Cardozo?' smiled Esmé.

'I am Cardozo!' announced the apparition. 'I wrote to you about a caravan.'

'Ah, yes, indeed. This here very van is for disposal.'

And Mr Cardozo came slowly down the steps to clasp us by the hand. He seemed to be conferring a knighthood.

'Enter, my dear,' he said to Esmé, and bowed both of us up the steps. The van was palatial in comparison with the Moggridge's. The Hostess was chromium-plated, and above its tiny mantel was a bevelled mirror. Two chromium lamps swung on brackets at either side of the mirror. The locker-tops were upholstered in crimson plush, and the double bunk folded up flush with the end wall. The length of the van was fourteen feet.

Mr Cardozo gave us a few moments to drink in this magnif-icence. Then he intoned, 'You need not be ashamed of entering upon no field in this here van. It will bear comparison in any company whatsoever.'

He led me outside to inspect the underworks and the belly-box. I pointed out that the roof seemed a little warped.

'That is not anything,' cried Mr Cardozo. 'We showmen re-canvas our tops yearly.'

I asked how this was done, and he explained that a strip of canvas was pressed flat on to a coat of thick paint, so that it stuck, and added yet another layer of protection against the weather.

'And my roof', he said proudly, 'is a genuine Mollycroft.'

I understood then that this was the term for a lantern roof. And when we went inside again he demonstrated how each little roof light opened for ventilation.

'You must drink tea with me,' said Mr Cardozo, and before we could stop him he went out on to the platform and boomed, 'Elizabeth!' A girl of about twenty-five came out of one of the buses which we had thought unoccupied. She was tall and had a beautiful figure. Her handsome features were as strongly marked as Cardozo's. Her hair was raven black, and her creamy pale skin was a little spoiled perhaps by too much make-up.

'My daughter!' said Mr Cardozo, presenting her with a grand air.

The girl moved gracefully about the van to set out the tea things, but from the way Mr Cardozo's eyes followed her she was most certainly not his daughter. Esmé explained about her glass of water, and over tea the old man divulged, to her delight, that he had been a lion-tamer.

'You must show the brutes who is master,' said he. 'I always remained on top, and consequent had no trouble throughout my career.'

I wondered how the girl fared with him. She was strange and quiet in a world of her own. I caught her gazing at Esmé with an indefinable look of wonderment, or perhaps appeal. In me she took no interest, as if she had been forced to drain the dregs of masculinity and found the lingering taste bitter. We had stumbled on some tragedy, and I watched for a sign to tell me that knight-errantry would be acceptable. I think the girl would have spoken to Esmé if she had had the opportunity, but the old man was too used to the centre of the arena to allow Esmé's attention to stray from him. At last I broke in on his monologue.

'We like the van very much, Mr Cardozo. But forty-five pounds is a little more than we can afford to pay. We have promised to see one other van in Birmingham, so in any case we can't decide at once.'

'You tell me where this here van lies, and I'll inform you of its condition. There is not scarcely a van whose history I do not know.'

But Esmé became at once secretive. 'I forget the address. It's in the car,' she said.

'I will escort you to this here car,' announced Mr Cardozo. We climbed out of the van. The girl murmured good-bye as we ducked through the postern gate. I stole a last look at her, but she was walking slowly, head bent, towards the converted bus. She never gave a backward glance. Mr Cardozo took in the points of the Bentley, while Esmé pretended a search for the address, which she had no intention of disclosing.

'I must have left the letter at the hotel,' she exclaimed at last, with the air of an innocent child. Then: 'Goodbye, Mr Cardozo. We'll write or call back about your van without fail.' The lion-tamer bent low over her hand, clasped mine, and stood in an attitude as we drove away.

'A nice van,' said Esmé, 'but I think we can find a better at the money.'

The last of the Birmingham caravans was only a few streets away. It was again parked on a vacant lot shut in by a high hoarding. As we approached on foot, Esmé scanned the owner's letter.

'This one belongs to a woman,' she told me. 'It sounds very nice. It's a Mollycroft, with a Hostess stove, belly-box and blooms. She's asking forty pounds.'

'I'm sure the blooms alone are worth that,' I whispered, as we entered the lot.

There were several vans and converted buses within. Right next to the gate was a very smart maroon van, with cream underworks and roof. On the shut bottom half of the door was the name A. Finney in chromium letters.

'That's the one,' said Esmé. But before we could approach, a female voice came from one of the buses.

'If you wants Annie go in an' wait. She's on an errand,' said the disembodied voice.

We did not quite like to make ourselves so at home in a stranger's van, and hesitated.

'Go on up!' ordered the voice imperatively. 'She expects yer. Go on! She won't eat yer!'

So we went up, and entered an extremely clean and tidy twelve-foot van. Its design inside was typical of all the others we had seen. But the stove was more heavily chromed than even Mr Cardozo's, the mantel mirror was of thick plate-glass, the swinging lamps were gleaming, and mica hats sat on top of their globes to save the ceiling from smoke. The double bunk formed a pleasant daytime divan. I knew at once that we would buy the van. And at that moment of intuition the yard gate banged, and a moment later a round red face appeared at floor-level, and began to climb slowly upward like a rising sun as some hidden means of propulsion thrust it up the ladder. The rotund face was set direct on to a larger sphere, which was draped to floor-level in an amorphous black gown. No feet were visible, and I surmised that the sphere was set on castors as it slid towards us.

'I'm Annie Finney,' said the face. 'Hup! Pardon, I'm sure!' An aura of the taproom welled slowly from the smaller globe and settled like a pall upon us.

'We were told to come straight in— ' began Esmé.

'That's right!' beamed Mrs Finney. 'You tell yer young man to look at the underworks, while I shows yer wot's wot in 'ere.'

Thus exhorted, I went outside and found the van in fine order. It was mounted on pneumatic wheels, of which the front pair swivelled to the pull of the drawbar. The bellybox was fitted with a neat lock, and the Mollycroft roof was sound and tight. When I returned inside, Mrs Finney immediately fixed me with a glassy stare.

'Did yer see the blooms?' she whispered, as if she were an Israelite speaking of the Ark of the Covenant. 'Did yer see them blooms? Cost me every penny of fifteen pun to 'ave 'em fitted, strike me dead if I tell yer a lie.' She paused, but there was no immediate heavenly manifestation. Reassured, she went on: 'There's years of wear in them blooms. Yer can travel the length and breadth of the country before they'll be wore out. And the van rides that comfortable...'

Blooms! Blooms! Balloons! The famous blooms were, of course, none other than balloon tyres. I glanced at Esmé, who avoided my eye. But Mrs Finney was in full discourse again.

'Yer must drink a cup of tea, buy or no buy,' she insisted, and became thunderstruck when Esmé declined. However, I drank tea, and Esmé sipped water. At last Esmé said, 'It's a very nice van, but I've promised to see one more at Burnley before I make a choice.'

'You see it, dearie. But mind the ticks! There's no bugs in Annie Finney's van, but there's some wot swarms in 'em.'

Esmé promised to keep an eye open for livestock, and to write to Mrs Finney, buy or no buy. We drove off once more on our pilgrimage, and I think it was the dreariest journey I ever made. It was midday when we left Birmingham, and low grey clouds vented an intermittent drizzle. A foggy haze hung yellow over the desolate, semi-derelict country as we drove north through the sickbed of industry. Groups of broken men, hunched against the damp, huddled disconsolate as they strove to strike a spark to warm the raw atmosphere of the English Sunday. We passed through Wolverhampton, Stoke, Stockport, Manchester, Bolton, Bury. Each larger town was linked to the next by a spawn of bleak small towns. We drove through a hundred miles of prison, where all the inmates were lifers. Burnley was more miserable than most of these towns, and we finally found the van on the outskirts, standing stark in a bastard half-world which was neither town nor country.

'They're only asking fifteen pounds,' said Esmé. But as soon as we approached we saw that the van was no use to us. It was very old and had been cheaply made. The wheels, which had iron tyres, were removed and stored underneath, and the under-carriage was propped up on four posts. A pathetic attempt had been made to establish a few plants about the van, and their sooty leaves drooped despairingly under the drizzle. A man and a woman came to the door at our footsteps. He was consumptive, lame, small and undernourished. She was small too, and wiry. She obviously lived on her nervous energy, and did not break down because she dare not. Their name was Wilkinson. They showed us over the van. A fresh baking was on the table, and Mrs Wilkinson praised her oven with shabby pride.

'We have to move', explained Mr Wilkinson, 'because the sanitary authorities won't let us stay any longer. But we can't go into a council house. Where's the rent to come from? I've only me dole. Me leg went bad after the War, and a cripple can't get work. And they won't let me have a pension.'

'You see,' broke in Mrs Wilkinson, 'if we could get a few pounds for the van we might get into a council house and take lodgers.'

But I knew these two were doomed, slipping faster and faster down the crumbling pit dug by civilization to catch the sick, the failures, the unwary. We left Burnley feeling very depressed. Our wild, clean valley was a goal more than ever to be desired. Why should some of God's creatures be condemned to work in pits, in the pandemonium of factories, while others such as we, had clean rain, clear sun, fresh winds, to temper mind and body?

We drove up our valley at last as the moon came over Siabod. Stars winked in and out as drifts of cloud played idly over them. The moon lit a shimmering broad track across the lakes, and the river became a luminous ribbon. The beauty took our breath away, as it does each time afresh. But we thought of a hundred miles of raped country.

'Man's inhumanity to man,' I murmured. And then Luck and Mot heard the car and came tearing down the hillside, over the wall and along the road to meet us. We stopped to pick them up, and suffered gladly their wet embraces as we stormed up Dyffryn hill.

'But perhaps few townspeople would like to live here,' said Esmé shrewdly, as later we stood a moment on the terrace before going to bed. The valley was a Paradise which glowed with a fairy light. Its colours were silver and black.

'That's the tragedy of it,' I said. 'We have lost our sense of values. We can appreciate nothing but the concrete, value nothing but the machine. Machinery has killed the soul of man, instead of liberating it. Oh, come on in! I'm going to bed.'

Quite soon Esmé wrote to each of her caravan people to say that she did not think their van just suitable for her purpose. But to Mrs Finney she made an offer of thirty-five pounds. In due course this was accepted, with a proviso that the sum should be paid in cash.

'How on earth will we get the van home?' asked Esmé, a little aghast at the magnitude of her enterprise. I felt guilty about the expensive bills I had caused her on our trip.

'I'll fetch it with the car,' I promised.

I took the Bentley to a blacksmith and had him fit a stout draw-bar. And one Sunday Esmé and I again drove to Birmingham, with Thomas as a grinning, wind-swept passenger in the back seat. We reached Mrs Finney's yard at last, and when we all climbed out of the car I saw that Thomas was bleeding from a wound on the bridge of his nose.

'I did put my head over the side to see for the wheel turning,' he explained, 'and you did go over a bump, and the side come up and hit me on my nose.'

Esmé was in close converse with Mrs Finney, whose round, portentous face was between mirth and tears. Esmé handed her the jealously guarded wad of banknotes.

'Thank you, dearie! No, I don't need to count them if you say so. Oh, dear! However I'm going to put up with livin' in one of them buses and humbugging myself I'm sure I don't know. You got a luvly van as anyone may be proud of dearie. You can go in any company in that van, and she draws light as a fevver on 'er blooms.'

While the two conversed, Thomas and I dragged the van out into the road, where we hitched the tow-bar to the new drawbar of the Bentley. The caravan weighed a ton and a half, and the big car looked like a tug hitched to a towering liner. My heart misgave me a little.

By the front platform of the van was a small hand-wheel. When screwed down, this clamped a set of wooden brake-shoes on to the blooms. The law says that any trailer over three hundredweights unladen must be equipped with brakes, so it was Thomas's duty to travel in the van and to satisfy the letter of the law by being on the spot to operate his little hand-wheel. We took another look at the car and van, and even Thomas became subdued.

'Come on, Thomas!' I said. 'We'll go and have a drink.'

We did. When we returned, Mrs Finney embraced Esmé warmly, and gave an excited cheer as her van swayed lurching away behind us. Whenever I glanced back, the monster appeared about to overrun Esmé and me, and Thomas's apprehensive face stuck through the open half of the door like a figurehead. Our Juggernaut progress astonished Birmingham. Policemen felt for their whistles on principle, because they had seen nothing like this before. But no one stopped us. I think our slow progress looked inevitable, as if governed by supernatural forces. And once in the open country we gained confidence. The speed increased to a thundering twenty-five miles an hour. Esmé stole proud glances at her purchase, and Thomas disappeared within to read the Sunday newspapers.

We stopped at an inn for lunch. From somewhere Thomas acquired half a bucket of coal. Soon after we had restarted, the chimney of the van was smoking merrily, and Thomas presently appeared leaning over the half-door with a cup of tea in his hand. Darkness overtook us on the high moorland road by Cerrig-y-Drudion, but Thomas lit the lamps in the van. Many a Welshman hurrying to chapel must have felt his flesh creep at the sight of the devil running away with a cottage. I expect our naming, sparking passage has melted into legend.

But we reached Dyffryn in the end. Thomas came out of the van and said the trip had been a treat. His nose was now much swollen. John Davies came hurrying from the cottages.

'By damn!' he exclaimed, after he had peered into the cosy caravan. 'Them shypsies do look after number one!'

18

The Dipping

THE LATE SUMMER IS CONTINUOUSLY BUSY with shearing, hay and snack bar, but there is also the dipping to complicate further the disposition of our waking hours. We dip twice. The second dipping takes place at not less than ten days after the first and at not more than a fortnight. Dipping is usually supervised by the police, but the control is exercised separately by each county. It is carried out to combat a disease known as 'sheep scab' and if an area has long been free of this trouble the authorities sometimes leave it to the farmer to decide whether he will dip or not. The north part of our county has not been under compulsion for the past two years, but we and our neighbours still carry out the dipping religiously, because in hill districts prevention is much easier than cure. I have never seen scab myself, and hope that I never shall, but before compulsory dipping the disease was very prevalent. It is called 'scab' because of the appearance of the wound, which is caused by a swarm of infinitesimal mites. These mites are so irritant that they speedily cause a sheep to lose condition and finally to die. And the trouble is extremely infectious, because rubbing-places harbour and soon pass on the mite, as does chance contact. Most scab dips are arsenical. The poison is mixed with a bulk of soapy material in liquid, powder or paste form, which when diluted with water impregnates the fleece down to the skin, and renders invasion fatal to the scab mites for a period, theoretically, of twelve months. The proportion of arsenic must be fairly high. A forestry man once told me that young trees will die from the effect of the arsenic in wool which has been left twined about their twigs by trespassing sheep.

Our first complete gathering of the mountain after lambing time is carried out now. It is late August, and the lambs are ready to be weaned, so that it does not matter if they do lose their mothers for good in the press. We gather on the day before the dipping, because we shall have a very full day's work on the morrow. The mountain is very crowded, with perhaps three thousand head of sheep, and before we have moved a mile from the eastern boundary the animals are thick before the dogs towards the lower end of the line. Up high, as always, John Davies and I see fewer sheep, and it is difficult to believe that there are more than a few hundred upon the mountain. But as we turn the Glyder and plunge down towards Cwmffynnon the flock seems immense, used as we have been to seeing it only in two halves. We turn at last into the road for the last part of the way to the pens, in order to avoid the enclosures around Dyffryn cottages and house. The bobbing serpent of sheep fills half a mile of highway. Each man walks behind a section the better to hasten on the sheep, for soon fifty cars or more are clamouring behind us, and as many are slowly filtering through from the front. Some of the motorists shake clenched fists at us. When they are asked to whom the sheep belong, the men forget their English and give an impassioned harangue in Welsh. Other motorists jump out with cameras and ask eager questions, which receive most inaccurate replies. To these Davydd Pentref is usually passed off as the owner, and the motorists say, 'Oh, what a little man to have so many sheep!' And that is even after Davydd has swollen an inch or two with pride.

At last we turn the head of the column into the field below the house. Presently the last stragglers are hustled through the gate, the dammed traffic flows away in spate, and we men climb up the hill to see what food Esmé has set out for us. And afterwards the gatherers go their ways with a promise to be early at the pens next morning.

If the dipping day is rainy we postpone the operation now that the work is optional for it is useless to dip wet sheep. The saturated fleece soaks up but little dip, and the falling rain soon washes out that little without giving it time to harden and set in the fleece. But in the days of compulsory dipping the attendance of the policeman had to be booked some days previously, and then it was not so easy to put off the day. Esmé and I pass a restless night by reason of the noisy flock under our windows, and I am glad when it is time to get up, have a quiet breakfast and to call Luck and Mot. John Davies and I have a telepathic communication which is really rather odd. Without knowing in the least where he is on our four square miles of land I can be sure of walking to cross his path. I cannot explain why. And on such a day as this, as I reach the brow of the hill below the house, Davies is standing on the road at the bottom, directing Bett and Jim up the steep field. Without instructions the impetuous Mot races to meet the other dogs, jealous of his share of sheep. He can do no harm, and I leave him unchecked to his excess of zeal. But the more disciplined Luck looks down his nose at this exuberance, and becomes even more than usually controlled in his neat actions, in order to emphasise to me the difference. Thomas has had milking to do, and joins us by the pens, and the three of us stand behind the flock as it passes interminably through the gate into the pen enclosure. The dogs, tongues lolling, watch like lynxes, and harry excited lambs who try to break back. Gwylim Pantffynnon, Davydd Pentref, Bob Henblas and several others arrive in ones and twos, and with their help we turn the sheep into the pens proper.

Each year at this time I realise with a shock that my annual sale is only about three weeks distant. I wish to sell all the four-year-old ewes and all that season's wether lambs. It is impossible in practice to gather clean our whole flock. At the first gathering for dipping, perhaps two thousand five hundred sheep may come in out of three thousand, and the numbers may be no higher at the gathering for the second dipping, but many of the individuals will be different. Thus, by picking through the flock for four-year-olds and wether lambs on each of the two dipping days, we are reasonably sure that nearly every sheep has been through our hands on the one day or the other. So we fill the sorting pen. First we catch out all the wether lambs and turn them into an empty pen which leads off the sorting pen. Then the men form a line across the narrow pen and press slowly through the ewes. They examine the mouth of every one, and the four-year-olds are turned down with the wether lambs. Then the residue of the sorting pen is released again to the big enclosure which surrounds the pens. Again and again we refill the sorting pen. I usually have to argue a little on these picking days. The reason goes back a long time. Welsh sheep-farmers will forgive a ewe anything if she is young. They will look at a miserable type of yearling and say, 'She have plenty of time for alter.'

But a soft fleece will never become weather-resistant, nor will poor bone become sturdy, nor stunted growth generous. Naturally these poor types of sheep are responsible for a high proportion of the winter mortality, and they transmit much of their deficiency to their lambs. The old-time farmers used to admit in a backhanded way that it was poor policy to keep this type of young ewe, because they used to send them out to winter just like ewe lambs. My prede-cessor at Dyffryn used to winter away about two hundred ewes at a cost of about eight shillings each. But with the help of John Davies, who surprisingly agrees with me, we always pick out this class of ewe for sale, irrespective of her age. At first the culling was heavy, but after a very few years the policy of ridding the flock of these weaklings took effect. And now we do not winter any ewes away, and buyers at our sale appreciate this point. For if they buy ewes which have never had the advantage of lowland wintering, save as lambs, there is obviously much more room for improvement in the stock.

The dipping bath is about twenty feet long, and is too narrow to allow a sheep to turn round. At the far end of the bath, steps lead to a concreted dripping pen where about a hundred sheep can congregate to shake themselves dry. The surplus dip runs off them and back into the bath again. The bath holds four hundred gallons of dip. A hosepipe is laid from a nearby stream, and several ten-gallon drums are kept ready filled with water to replenish the dip which will be soaked up by the sheep. While the rest of us sort, Thomas mixes the dip to pour into the bath, which is ready filled with water. The dip comes to us in hundred-pound drums, and each drumful costs about seventy shillings.

One pound of the paste is sufficient to make six gallons of dip, and Thomas has spring scales to help him work out his proportions. He mixes a little paste with water in an empty drum to make the fluid, then pours it into the bath, until at last the strength is correct. There is a strong, clean disinfectant tang in the air. And the sheep which we have loosed into the big enclosure smell what is to happen to them, and are reluctant to be driven back into the pens again.

We deal first with the sale sheep, while the dip is new and clean, for a yellow bloom is mixed with it which will colour the sheep attractively. We drive about fifty sheep into the small pen at the head of the bath, and Pricey Garthllugydadaith, who is always overflowing with energy, takes the first turn. Old Bob Henblas, who has outgrown such frolics, stands beside the bath with a long-handled pole whose iron head is shaped like a concave crook. As Pricey slides the first sheep into the dip, Bob presses the curved iron on her neck and ducks her vigorously. The ewe swims frantically along the narrow passage and scrambles up the steps into the concrete pen. Her legs give, and she staggers under the weight of the liquid in her coat. Pricey works as if he is out to win a bet, and he swings the sheep so rapidly over the edge of the bath that old Bob, who is being splashed from head to foot, becomes angry.

'There be plenty of time, boy!' he reproves, in his slow way.

'You must give to me a chancet to push them down!'

Pricey finishes his pen, and I take the next turn. The recently shorn sheep are difficult to catch and to pick up. I can only manage to lift them by sliding one arm about their necks and by grasping the loose skin of the hind flank with the other hand. And now and again a ewe kicks violently as she is tumbled into the bath, so that one has to take good care of one's groin. We work fast while we are still fresh, but the labour is severe, for the sheep smell the dip and struggle against us from first to last. Each time the dripping pen fills up, we pause a moment until the stream of returning dip dwindles to a trickle, then I go round to open the gate. The yellow-coloured sheep dash free into the big enclosure, and stand about shaking themselves and blinking their eyes. I count each penful. By midday we can just finish the sale sheep. The count tells that seven hundred of them have come in, and this is quite satisfactory for the first gathering. We turn these ewes and lambs together into the ffridd. Some few of the wether lambs have four-year-old mothers, and trot away contentedly. But the bulk of both ewes and lambs is disconsolate, and the sheep range bleating about outside the wall of the big enclosure. We leave Thomas to replenish the bath, which has lost about a hundred gallons of liquid, and he begins to mix sufficient fresh dip to balance the added water. The rest of us, splashed, filthy and smelling of chemical, climb up to the old farmhouse and sit down to lunch. We feel the worse for Esmé's daintiness.

As we return to the pens we meet Thomas coming up for his

belated meal. His belly is indignant at the delay, and he says as he passes us, 'Indeed, I do have a pain in my guts, something terrible.'

But the men are more cheerful now. As Gwylim Pantffynnon takes his turn at dipping, the others contrive to introduce secretly a steady stream of sheep into the pen behind his back. Old Bob Henblas with his pole pretends to become indignant, and calls on Gwylim to hurry up.

'You are the most slowest on the lot of them!' says Bob.

'His wife be killing him,' explains John Davies. 'He have never got over making that baby.'

Poor Gwylim sweats and struggles, until suddenly he becomes suspicious and swings round in time to see a bunch of sheep being driven furtively into his pen. Before he can attack anyone we swamp him in a flood of sheep and lambs, and I jump in among them to take my turn. The sheep are growing heavy now, and it is a relief to grab a lamb and to lift it by the comfortably long wool. But presently John Davies returns from an inspection of the back pens.

'By damn,' he says, 'the flamers be growing less!'

And when Esmé comes down with tea and sandwiches and buttered scones there will be only a few hundred sheep left to do. We usually finish a couple of hours after tea, and the helpers hurry away to their farms to help cart the evening hay. But John Davies and I have to drive the flock, diminished by the number of the sale sheep, back to the mountain. When I return to the house for a bath, the dirt which washes off me nearly blocks the drain-hole.

'How many did you dip?' Esmé used to ask me in the early days at Dyffryn, when I had come downstairs again clean.

'Seven hundred sale sheep,' I would perhaps tell her, 'and sixteen hundred of the mountain flock.'

Esmé would look startled.

'That's only two thousand three hundred. We're seven hundred short.'

'Yes, but the east wind we've been having will have driven hundreds away westward into Cwmffynnon hollow and along the ridge towards the big Glyder. The wind will change by the time we gather for the second dipping, and the rest will come in.'

And they do. Ten days or a fortnight later we gather again. If the usual westerly weather is prevailing, hundreds of sheep whose lack of yellow bloom proclaims them undipped are mixed with the coloured bulk of the flock. The bleating sheep give Esmé and me another sleepless night as they range about beneath the house. Next day we again go through the routine of picking wether lambs and four-year-old ewes. We glean another two hundred. And this time, too, we pick out all the ewe lambs and pen them by themselves. For it is now early September, and on the first day of October these ewe lambs are due to travel away to the mild Conway valley to winter. The dip is at half-strength for the second dipping. A pound of the paste makes twelve gallons of fluid. Now that the work is not compulsory we do not bother to dip again the flock of sale sheep. For the wether lambs will mostly be eaten by Christmas, and the ewes will be purchased by a score of small lowland farmers who can give them almost individual supervision. But we take no chance with our mountain flock, and dip them again thoroughly. For should trouble ever start it might be weeks before we would notice it, and afterwards the authorities would insist on many double dippings each year until the scab was stamped out. The cost of money, time and trouble cannot be contemplated, so all we mountain farms are conscientious in prevention,

When dipping becomes compulsory – which it does spasmodically in various counties owing to outbreaks of sheep scab – the work is taken very seriously. The policeman who has to be present supervises carefully the mixing of the dip in the correct proportions. He watches to see that the job is not scamped, and he counts the number of sheep dipped and gives a certificate of dipping on which the number is stated. It is then necessary to have a movement licence in order to send sheep into another county. And the neighbouring authority has to be satisfied that all the stock has been thoroughly dipped. Some short-sighted farmers grumble at these restrictions, and at other restrictions too, which affect different types of beasts, but most of the veterinary types of order are sensible, and are founded on necessity. We are, it is true, in the gravest danger of transferring our national occupation from shopkeeping to bureaucracy, but in the respect of such orders as the dipping, warble-fly and foot-and-mouth regulations there is no doubt but that great progress has been made in the lessening of disease. Most farmers will agree. But in other aspects farming, above all trades, is not readily susceptible to central control. Each district has conditions basically different from those of adjacent areas. And in every district each farm requires different treatment, and in each farm each field. It is impossible satisfactorily to appoint commissions to deal wholesale with fertilizing, marketing and crop growing. And there is danger that these commissions will be appointed. Each farm would need an individual Civil Servant to treat its peculiar problems. Too soon we should have a bureaucrat to every farmer. The trend is all too patent in every direction, for productive workers decline in numbers as the largely parasitic governing departments grow. Let us fight bitterly for freedom of individual effort.

19

The Record Walk

DURING THIS SAME LATE SUMMER SEASON two years ago, Esmé and I, with mixed feelings, found ourselves front-page news in the daily papers. In Wales there is a famous walk known as the 'Three Thousands'. The Three Thousands are the fourteen named peaks in Wales which exceed three thousand feet in height. A great many years ago it used to be the mountain walker's ambition to ascend these peaks in succession without pause for sleep or for a long rest. For it is not very long since humans began to regard mountains as a proper sphere in which to move. There are some surprisingly recent accounts of ascents of Snowdon in which the most hair-raising circumstantial accounts are given of eagles, spirits, nose-bleeding and lack of air. And Munchausen's stories would have been considered conservative by some of the perpetrators of early Alpine writings. But with man's increasing familiarity with his new recreation the circuit of the Three Thousands in Wales was brought within the time limit of twenty-four hours, and later, quite recently, to twelve and a half hours.

In the spring of 1938 a friend was staying with us. Capel and I were sitting one clear, frosty afternoon on the summit of Glyder Fach. The view was splendid. Snowdon, Crib Goch and Crib-y-Ddysgyl rose beyond the huge hollow of Cwm-ffynnon south-west of us. Our own range stretched north-west

– Glyder Fawr, Y Garn and, four or five miles away, the lonely outpost of Elidyr. North-north-east Tryfan stabbed skyward. Llyn Ogwen lay at its foot like a careless jewel. Half a mile beyond the far shore of the lake was another three-thousand-foot hill, Pen-yr-Oleu Wen, a precipitous tangled slope of boulders and heather. Past it agai were the rounded humps of Carnedd Dafydd and Carnedd Llewelyn, hills almost as high as Snowdon herself. And hidden behind their swelling crests were the last of the Three Thousands, softer grassy summits which overlooked the Bay of Conway. Capel and I looked for a long time at the glorious panorama of mountains, and I think that the idea came simultaneously to us to make an attempt on the record for the Three Thousands. We planned our route there and then.

'We must start from the top of Snowdon,' said Capel, 'and do the three Three Thousands in that group first. Then we'll drop down from Crib Goch to Pen-y-Pass and slant upwards to Elidyr. From there we must work back over the summits to here, and from here over Tryfan, down to Ogwen, and on up over the Carnedds.'

'The route shows about twenty-four miles on the map,' I said. 'If we take ups and downs into account we'll have to cover about thirty.'

We had a long look at the map.

'Our total descents will be about nine thousand feet,' I calculated, 'and ascents about eight thousand five hundred. To be exact, there will be four hundred and sixty-nine feet difference – the difference between the height of our first mountain, Snowdon, and our last, Foel Fras.'

When later on we explained to Thomas, who was most interested, why ascents did not exactly equal descents he was quite scathing.

'By damn,' he exclaimed, 'there do be nothing in it! It be all downhill!'

When we returned to the house that day Esmé was very thrilled with the plan. She insisted on her inclusion in the party, and promised to drop behind if we found she was holding us up. Capel was only making a short stay then, but he promised to return later in the year, and to bring a friend with him who was also interested in attempting the Three Thousands record. Capel proposed that he and his friend should stay for some weeks at the Climbers' Club Hut near Tryfan in order to train and to join Esmé and me in finding the best route. For at that time it seemed to us that it would indeed prove difficult to better the existing record of twelve and a half hours.

Esmé and I were, as always, very busy through the summer, but we found time to take several busman's holidays walking in the hills. We faithfully reported by letter to Capel any discoveries about routes and times. The walk lay in three distinct sections. The Snowdon group of three peaks stood together, joined by high ridges, which enabled one to reach each summit without serious descent or ascent. But from the third peak, Crib Goch, was a very steep fall to the top of the Llanberis Pass. The pass isolated Snowdon and her retainers from the group consisting of distant Elidyr, Y Garn and the two Glyders. We had here to lose two thousand feet of height, before we began our long slanting ascent of five miles or so to Elidyr. This stretch of country was broken and difficult. Our first attempt took Esmé and me two hours and forty minutes from Pen-y-Pass to the summit. There were large areas of rock debris which could not be crossed quickly. Often I was doubtful whether it was better to cross valleys direct in order to shorten distance, or whether it was better to contour round the heads of them, thereby conserving energy. Halfway was the huge headland of Esgair Felen, which jutted fiercely over the lower reaches of the pass. We had to cross its ridge at two thousand seven hundred feet, only to slide down a steep scree on the far side to the two-thousand level. One day we tried to find a way round the steep face which frowned on the pass. The place was forbidding, overlooking the narrow, grim and sunless ravine. The thread of road twisted and turned fifteen hundred feet below us, as if it went in fear of the stooping cliffs. But the angle of the ground steepened as Esmé and I worked under the main precipice. We are both rock climbers, and quickly knew that, roped or unroped, no sane people would have tackled that path. The rock was rotten, and no foothold or handhold was secure. We retraced our steps.

There was, as we then thought, no possibility of mistake in the Snowdon group. The summit of Snowdon is named Y Wyddfa. Y Wyddfa was only ten minutes from Crib-y-Ddysgyl. And Crib-y-Ddysgyl was connected by a knife-edge rock-ridge to Crib Goch, a mile away. Deviation on the Crib Goch ridge would have been misguided and unfortunate, since the rocks fall away from the crest on either hand at a lethal angle. From the tip of Crib Goch the summit of the pass lay visible beneath us. And again when we had made the best of our clumsy route from Pen-y-Pass over Esgair Felen to Elidyr, we could not go far wrong on the return journey over Y Garn and the two Glyders, since by bearing to the left at any time over the four or five miles between Elidyr and Glyder Fach we would come to the brink of the precipices which fringe almost the whole north side of the range. For we were worried about mist. The mists are sudden and impenetrable about Dyffryn. It would have been heartbreaking to lose the way during the attempt. But on these two groups, at any rate, the precipitous ground marked the way well. It was a physical impossibility to go wrong if one used the cliffs as a guide.

But the Carnedds were different. It was comparatively easy to find Pen-yr-Oleu Wen, Carnedd Dafydd and Carnedd Llewelyn in mist, because here again were connecting ridges whose sides were often too steep to descend. But from Carnedd Llewelyn over

Foel Grach to the last peak, Foel Fras, stretched a high, featureless moorland of bog and peat, which was notorious for mist. We could do little beyond hope for clear weather. Esmé and I timed ourselves often over these three groups of hills. We found that we needed an hour from Snowdon summit to Pen-y-Pass, five hours from Pen-y-Pass out to Elidyr and back over the peaks to Glyder Fach. It took another hour from Glyder Fach over Tryfan and down to the road at Llyn Ogwen. And from Ogwen over the Carnedds to Foel Fras was another three hours and forty minutes. This gave us a total time of eleven hours and ten minutes, when three periods of ten minutes for rest and food were added. In theory we should then have beaten the record by fifty minutes. Of course, we had only pieced together our times over portions of the course. It was possible that when we came to cover the whole distance in one stretch fatigue would add to the time we should take on the Carnedds. But we decided to offset this by our increasing fitness.

August came at last, and with it Capel and his friend Rex arrived. We went at once into solemn conclave. We were all jubilant at the times which Esmé and I had noted. Capel began to pencil a schedule on the map. Suddenly he jumped. Without a word he pointed to an insignificant spur jutting out for a mile on the north-west side of Carnedd Llewelyn. At its far end we read, 'Yr Elen, 3151 feet.' Esmé and I were too abashed to say a word. We had missed Yr Elen through sheer carelessness. I had known perfectly well that there were fourteen Three Thousands, and had not checked their numbers. The fifty minutes' grace which I had thought won from the record were swallowed up. Secretly Esmé and I were not altogether sorry about this new development, for new interest was promised.

The four of us walked daily for several days, and continued to time ourselves over parts of the route. Capel and Rex, always fit, began to harden to even greater fitness. Esmé and I became tireless. Our times improved, but we found that the Carnedd group with the lost Yr Elen included took forty minutes longer. We had very little time in hand to provide for mishaps. But Esmé became thoughtful. We used to meet Capel and Rex after dinner each evening at an hotel in the village, where we would drink a frugal pint of beer and smoke the pipe or two to which we were limiting ourselves. Esmé took to brooding by herself while she pored over a map. One night she sprang her Idea.

'I'm sure we'd save time', she said suddenly, 'if we were to walk straight down the road from Pen-y-Pass to Nant Peris. We could climb Elidyr from Llyn Peris. We'd make good time down the road and avoid that awful grind over Esgair Felen and all that broken country.'

But we three men shouted her down. Llyn Peris was less than three hundred and fifty feet above sea-level, while Pen-y-Pass was nearly twelve hundred. We stoutly maintained that it was a hopeless principle to lose all that height.

'Well, I'm going to time myself to-morrow that way from Pen-y-Pass to Elidyr.'

'I'll go with you', I said, 'for the exercise.'

We did go next day. I groaned inwardly as we lost each valuable foot of height in our descent of the pass. Elidyr grew in stature as we went down towards her base. We turned off the road up the valley of the little Afon Dudonyn. We had covered the four miles in about fifty minutes. The way up the mountain became very steep, but it was grassy. Not until we were quite near the top did we have to negotiate the fields of boulders which were so numerous on the Esgair Felen route. We were eating our lunch by the cairn after climbing the two thousand seven hundred feet from the road in under an hour. We had saved nearly an hour over the old way. That night at the inn we toasted Esmé as an excuse to exceed our modest quota.

We were now becoming very fit. We had worked out a summit-to-summit schedule, and from this we started to clip off minutes which totted up to a respectable total. We began to aim at a time of eleven hours, although none of us knew how the later stages of the walk would be affected by fatigue. And just when we were most jubilant, Capel burst another bombshell. We had been at great pains to collect all reports of quick times for the circuit, and made the belated discovery that a member of a mountain-walking club had completed the Three Thousand in ten hours and twenty-nine minutes. This was positively the fastest time in which the walk had been done, but we were none the more cheerful for that. Again we pruned our schedule. We ran downhill where the ground would allow of it. Rex one day performed the startling feat of reaching the road at Ogwen from Adam and Eve on Tryfan in thirteen minutes. This is a distance of a mile and a quarter with a fall of over two thousand feet on the most broken and rugged mountain in Wales. Many walkers consider an hour a reasonable time to take. We used Rex as a pacemaker downhill after this performance, while I set the pace uphill. But we found that for all our efforts we had little margin over the record. Esmé became thoughtful again. After the interval of a day or two she sprang another Idea on us.

'Why can't we leave Crib Goch by the North Ridge and strike the pass halfway down?' she asked. 'As it is, we go away from Elidyr while we make for Pen-y-Pass.'

Again we were all sceptical. The North Ridge certainly pointed straight at Elidyr, and would save us a mile and a half of road, but it is a very sharp ridge indeed. If the main Crib Goch ridge is knife-edge, this is razor-edge. And the country below it is very rough, with an unpleasant gully alongside a waterfall on the direct route which we should have to take. But we tried next day. The ridge would have been tricky in a high wind, but we were able to move quite quickly along it. And below we found the going better than we had expected. Indeed, Rex ran without pause down to the road. The new way saved us twenty minutes, and we toasted Esmé again. The rest of us now began to pay minute attention to the route. Many small savings of time or of energy spread over the whole distance of maybe thirty miles made an appreciable total. Our schedule shrank below ten hours, and as we grew more and more like chamois I at least began to think of nine hours, and at last persuaded Capel to remodel our times on this basis. The pace had now grown too hot for Esmé. She decided that she would try to beat the old record, but that she would be about an hour behind us. We suggested, therefore, that she should start from Snowdon an hour before us, so that we should all finish at the same time. But since we were to travel separately I became worried. It is unwise to move at breakneck speed alone over our rough, rocky hills. I could think of no man who could keep pace with her without training, or perhaps even with it. But she suggested Thomas. Thomas, of course, is always fit, and would not be too proud to allow himself to be relegated to the slower party. He jumped at the idea of what he called 'a real treat'. We took him with us on all our later expeditions, and he speedily put an edge on his condition.

At last we fixed the day. Capel made arrangements with the Snowdon Mountain Railway Company, who foresaw publicity, and offered to run a special train to reach the summit before 8am, which was the time at which Esmé wanted to start. And the publicity did come. Local reporters came first, and photographed a very uncooperative Esmé patting dogs, reading maps or gazing over the void in front of Dyffryn house. The big dailies followed suit, and we read many imaginative columns devoted to our doings. The day before the attempt we took things easy. Towards evening Capel drove his car to the coast near Aber, and turned up a road which led towards our finishing point of Foel Fras. He managed to reach within four miles of the summit before he had to stop. He left the car for our return next day. I had followed him in my car, and we drove back together. As we came up the Ogwen valley, dusk was falling. Tryfan reared her great bulk threateningly. Pen-yr-Oleu Wen was blackly forbidding.

'They've all got something up their sleeves,' I said.

And they had. When we tumbled out of bed at half-past five next morning a patchy mist was drifting slowly past Siabod. Capel and Rex arrived soon for breakfast. When the meal was over Siabod had disappeared. It was difficult to postpone the day. Rex's brother was meeting us on Glyder Each with food, and more friends were taking coffee and sandwiches to the top of Pen-yr-Oleu Wen. The Press, too, was represented at strategic points, and many acquaintances had told us that they were going to watch for us. And after our training it would have been a shattering anticlimax to wait when we were so keyed up.

'We must remember our public,' I said to Capel, who hates reporters.

We laced our boots and drove down the hill to collect Thomas, who was grinning with excitement. Quite a crowd was at the little terminus of the funicular railway at Llanberis. An early and daring news-hawk photographed us, and announced his intention of taking the train with us to obtain dramatic pictures of the start. The manager of the railway, shaved and wearing a bowler hat, gave us an official blessing, and was nice about the fare. Several quarrymen and railway employees raised a ragged cheer, the quivering engine gave a tremendous hoot, and we began to lurch up the steep track. Almost at once the mist swallowed us. The railway follows a steep ridge. Our reporter peered anxiously into the void and shuddered. We had no reason to doubt that his descriptions would be lurid. An hour later the little train ran level, as if suddenly dropping on all fours, and stopped with a triumphant screech. We disembarked stiffly from the cold open-sided trucks and walked, damp and miserably cold, into the summit hotel. It was a few minutes short of eight o'clock. Esmé and Thomas hurriedly gulped some hot coffee.

'Hope my leg will last out,' whispered Esmé to me. 'What's the matter with your leg?' I asked, startled. 'I strained a ligament the other day,' she answered. 'That's why I haven't been out walking for the last few days.'

And before I could make a protest or hear further details she and Thomas threw off their coats and vanished into the mist in the direction of the Crib ridge. The hotel staff and the engine-driver cheered excitedly, and the photographer fell over a rock while attempting to take a picture. Capel, Rex and I returned dispiritedly into the hotel, where we drank more coffee than was good for us, and beat our arms about to keep warm. The hour slowly went. We took off our coats and surrendered them to the engine-driver, who promised to deliver them later to us or to our next of kin.

At the stroke of nine we issued forth. The assemblage mustered several more unconvincing cheers, and the photographer bobbed about in a hide which he had found between two boulders a few yards along our route. We all ran. I was impetuous, and forgot the hours and the miles of rough travel ahead. Capel was more sensible, and prevented us from expending all our energy at once. Even so we reached Crib-y-Ddysgyl at a trot in eight minutes, and at once began a suicidal gallop along the Crib Goch ridge. The mist hemmed us close and concealed the pitfalls on either side. We scaled along the crest like steeplejacks on a rooftree. Every now and again the mist parted a little, and a clear funnel opened for a moment. I kept peering down, apprehensive of sighting sack-like objects lodged on a ledge or jammed between boulders. I knew that Esmé had a win-or-bust complex, and that Thomas was notoriously careless. But no signs of tragedy appeared during the brief spells of visibility. We reached Crib Goch nine minutes ahead of our

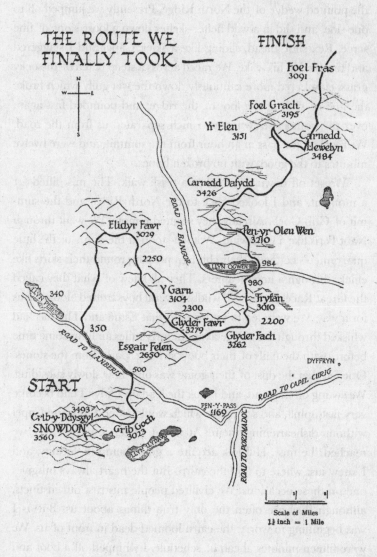

THE ROUTE WE
FINALLY TOOK —

FINISH

Foel Fras
3091

Foel Grach
3195

Yr Elen
3151

Carnedd
Llewelyn
3484

Carnedd Dafydd
3426

Elidyr Fawr
3029

2250

Pen-yr-Oleu Wen
3210

ROAD TO BANGOR

LLYN OGWEN   984

980

Y Garn
3104

2300

Tryfan
3010

LLYN IDWAL   340

350

ROAD TO LLANBERIS

500

Glyder Fawr
3279

Esgair Felen
2650

2200

Glyder Fach
3262

DYFFRYN

ROAD TO CAPEL CURIG

START

Crib-y-Ddysgyl
3493
SNOWDON
3560

Grib Goch
3013

PEN-Y-PASS
1169

LLYN LLYDAW

ROAD TO PORTMADOC

Scale of Miles
1½ inch = 1 Mile

nine-hour schedule. We did not pause, but angled out at once along the pointed wedge of the North Ridge. Presently we jumped off to one side, and slid in a wild helter-skelter down a loose slope of fine scree. Rex shot ahead, sliding like a skier. Capel and I staggered and tumbled in his wake. We raced across a steep patch of tussocky grass, clambered more cautiously down the wet gully which broke the line of cliffs at the foot of the ridge, and pounded fast again over the boulder-strewn grass which separated us from the road. We reached the pass in an hour from the summit, and were twelve minutes to the good, with no broken bones.

We set off down the road at a brisk walk. The mist lifted for a moment, and I looked back to the North Ridge and the summit of Crib Goch. Already they seemed remote. We went through Nant Peris like a whirlwind. Women stood at the doors of the little quarrymen's cottages, their children peeping round their skirts like chicks through a hen's feathers. They all knew of what they called the Great Race, and waved wildly. Several boys trotted alongside us for a way. We were led to understand that Esmé and Thomas had whisked through the village at about forty miles an hour some time before, with the nails of their boots striking sparks from the stones. One felt that the dust of their going was only now slowly subsiding. We swung off the road, and I took the lead up Elidyr. I had become very fast uphill, and set a pace which would extend Rex and Capel without disheartening them. At about fifteen hundred feet we reached the mist. Hill folk acquire a good bump of locality, and I knew just where to find the cairn. But the heart always misgives one on these occasions. We civilised people mistrust our instincts, although they are often the only true things about us. But as I was beginning to worry, the cairn loomed dead in front of us. We were fifteen minutes ahead of schedule. I whipped off a boot and slapped some cotton wool and plaster on a blistered heel, eating raisins at the same time. I was frightened of the heel. It takes very little to lame one, and the least limp would have ruled me out of the trip. But I never felt my blister again till the walk was over. Within a couple of minutes we were off again. We contoured around the head of the Dudonyn valley, which we had already crossed at a much lower level at the foot of Elidyr, and made along the ridge towards Y Garn. The mist lifted again. It was not yet midday, and our spirits rose, for there was still time for the weather to clear. We plugged steadily up the hard slope of Y Garn. From the summit I could see to the top of the cliffs below Glyder Fawr, but there was no sign of Esmé and Thomas. I was rather worried, because my eye was covering forty minutes' travel, and I thought that our exceptionally fast start would have brought us within sight of them. We plunged down into the hollow of Llyn-y-Cwn, and as we went the mist rolled back over the ground. It came seemingly without haste, but in reality as fast as a galloping horse. We had no time to orient ourselves. We thundered on at a trot. The time by which we should have reached the lake was past. Rex suggested turning left-handed. He was right. We had previously swung in a half-circle, and had been heading down the gentle grass slopes which led to the cliffs above the Llanberis Pass. But now we struck the lake and began to scramble up our well-known route to Glyder Fawr. As I was jumping and poising among the vast heap of rocks immediately under the summit I felt cramp in my calves and thighs, but along the comparatively level ridge between the two Glyders the feeling wore away. There was something uncanny about this short stretch of a little over a mile. Often we had timed ourselves between the two peaks. Sometimes we had hurried, sometimes we had been more leisurely, but always the time was eighteen minutes. Rex's brother answered our shouts as we approached Glyder Fach. In spite of our confusion in the mist near Llyn-y-Cwn, we were ahead of time. We wolfed down bread-and-jam sandwiches. Rex's brother told of a photographer who had insisted on viewing us in action. He had been sent towards Crib Goch, whence he had returned, shaken, to the bar of the Pen-y-Pass hotel. We rested for only five of the ten minutes which we had intended to allow ourselves.

The mist was thick. We navigated through the strange rock shapes of the little Glyder. On our right we passed the Cantilever – a huge long slab balanced over two uprights like a Druid menhir. On our left was the Mushroom Garden – a copse of queer huge stones standing like a clump of toadstools. We began to slide down a gully towards Tryfan. It was steep. Loose stones chased us. We reached the connecting ridge, and began our hard climb to the top. Presently Adam and Eve showed through the thinning mist. Just before we reached them we passed the Elephant. The Elephant is quite a small rock which leans rather wearily over the east face of Tryfan. He has considerable magical power. We were delighted with our progress, and touched him in turn.

'Thank you, Elephant,' said Rex. 'Thank you, Elephant,' said Capel. 'Thank you, Elephant,' said I.

The descent from Tryfan was Rex's province. After his thirteen-minute feat we wished to do no more than follow as best we might. The mist had risen, and from where we were we thought we could see knots of people on the Ogwen road. Rex led off at a tremendous pace. We followed down the loose, north-facing gully. I kept a hand behind me on the sharp-sloping ground. And halfway down we passed Esmé. She was limping. Her leg was hurting her badly on downhill going, but was not so bad on the level or when climbing. Thomas, cheery as ever, was encouraging her. We swept past and were soon on the road, where we shook off more reporters and some sightseers. We were twenty minutes ahead of our nine-hour schedule.

Pen-yr-Oleu Wen was a nightmare. I suppose we had by now covered more than twenty miles, and had done the bulk of our eighteen thousand feet or so of ascent and descent. We knew that we could finish if we could but gain the Carnedds, for there was only comparatively level ground up there. But the long climb over extremely rough boulder-strewn and heather-covered hill-side was heartbreaking. I led up as fast as I dared, and to our astonishment we took less time than we had ever recorded when untired. Friends were sitting on top with coffee and sandwiches. Unfortunately the milk in the coffee, shaken by the climbing, had curdled like junket. We paused for five minutes, and then jubilantly went off towards Carnedd Dafydd. But almost at once the mist closed down for the day.

The Carnedds are notorious for their mists. But at no time have I seen one so thick anywhere. We walked in Indian file, almost treading on one another's heels. The rear man could but dimly see the leader. At first it was not difficult to find the way.

It was all ridge country from Pen-yr-Oleu Wen, over Carnedd Dafydd, and along the spur to Yr Elen. I do not think that we wasted any time here. But when we left Yr Elen to strike upwards to Carnedd Llewelyn the task became more and more difficult. As we moved back along the crest of the spur I shouted at intervals. I thought it just possible that Esmé might have continued, and that if she had she might be making out towards Yr Elen. And to my amazement a faint 'Halloo!' came back from somewhere lower down the hillside.

'By God, they've made good time!' exclaimed Capel. But I was leading now, and my own problems beset me. I was afraid of missing the cairn of Carnedd Llewelyn and of wandering about lost in the featureless moorland beyond. It seemed foolish moving at top speed with no objective much beyond my nose,

'I've missed it!' I gasped miserably to Capel. For lack of better guidance we raced on. And there was the cairn! I almost ran into it. I slapped it with my hand and relinquished the lead to Rex. Capel and he between them were magnificent over the last three miles. Foel Grach and Foel Fras rise only a few feet above the high level of the country. We seemed to be surrounded by damp white cotton-wool, but I think they kept almost exactly to the most direct route. We found and passed Foel Grach. I was terrified lest at this last minute we should lose ourselves with the record so amply within reach. But presently we came on the boulders which mark the low summit of Foel Fras. We broke into a stumbling run and came on the cairn, a hollow stone structure like a grouse butt, into whose shelter we sank gratefully from the biting wind. It was not surprising that no one was there to greet us. Capel and Rex compared watches. We had taken eight hours and twenty-five minutes from the summit of Snowdon. We had beaten the record by two hours and four minutes. We sat shivering for a while. We were in shorts, and had one spare sweater each to go over our open shirts. But all our clothes were wet through. I set off back to look for Esmé. Quite soon she answered my shouts. I took her hand, and she limped at a round pace to the shelter. She had been nine hours and twenty-nine minutes on the way. This was exactly one hour less than the previous record by a man. Thomas was beaming with gratification.

Esmé was in considerable trouble with her leg, and the five cold, wet miles to the car must have been a nightmare to her. Emissaries of the BBC wished us to wait at Aber to record our experiences with a broadcasting van as soon as it should arrive. But, cold and wet as we were, we had ample excuse to evade waiting for the ceremony. We went straight back home for hot baths and a meal. Rex fell asleep after the soup. But Esmé scored best. The BBC invited her to London to be televised, and paid her expenses.

20

The Rock Climbers

ON THE HILLS THERE IS a most sharp distinction between walking and climbing. One never climbs Snowdon. The ascent of Snowdon by however steep a route is a walk – unless, of course, one reaches the summit via the nine-hundred-foot precipice of Lliwedd, which is quite properly a climb. Climbers are very conservative folk, and in not a little danger of being pompous. There was considerable gentlemanly disapproval expressed for our record walk and for the publicity, unwanted though it was, which had attended it. The climber has long since mastered the use of understatement as a medium for exaggeration. T. E. Lawrence employs this method in his *Seven Pillars of Wisdom.*

Climbers' organizations keep voluminous records of the exploits of their members. All these accounts are toned *diminuendo*, but the reader sees the writing between the lines, as he is meant to do. The newspapers captioned our walk in the language of journalese, and the tale lost nothing by their telling. We liked the result no better than did the climbers.

If a party of climbers were to get into grave difficulties during a gale on some hair-raising pitch of a thousand-foot cliff, and if they escaped with their lives, the leader would write up the account something as follows:

On September 8th, I led a party up Hell's Punchbowl. Tumbler climbed second, and Excelsior-Jones third. Owing to the length of the pitches we used two ropes of a hundred and twenty feet in half-weight. We wished to decide whether a variant which lies to the east of the ordinary route was practicable. There are no holds on the first pitch of a hundred feet. Progress is made by jamming the fist in a crack and proceeding thus hand over hand. At the end of this pitch is an ample stance the size of a saucer. I belayed Tumbler's rope over a knob of rock, and shared my stance with him when he arrived. I then moved upward over a vertical slab of sound rock, which was distinguished by its paucity of holds. However, there are several large fissures at convenient intervals in which the climber may place the top joint of the little finger. I took out the rope to its full length before finding a suitable belay in two huge flakes of rock which sprang from the slab. I was able to insert a boot-nail between the lower flake and the slab, while there was room to slip Tumbler's rope behind the upper flake. Tumbler brought Excelsior-Jones up the first pitch, and left him comfortably ensconced, while he himself came up to me.

This sort of thing goes on for several hundred feet, until:

The face now took an angle away from me like the corner of a house. There were no holds. Progress was continued by embracing the angle with arms and legs. The members of the party lost sight of one another in mist, and it began to drizzle. The pressure of the water which was coursing down the rock-face occasionally swept an arm or a foot from its position. The pitch was somewhat exposed, and caution was necessary. At this juncture a pebble loosed by the rising breeze struck Excelsior-Jones a light blow on the shoulder, breaking his collarbone. He had a little difficulty in completing the climb, which we did in two hours and ten minutes. I understand that this interesting route has not before been attempted.

[Signed]   *Chomondeley Upjohn*
                    Thomas Tumbler
                    pp. Glyn Excelsior-Jones

Esmé and I took exception to the haughty attitude of the climbing fraternity. After all, we had chosen to cast our lot among the hills, so their moral reproof that our race over the summits showed no reverence for the hills did not hold good. And many climbers see little of the hills. Their physical and mental view is limited to a few feet of rock. A suggestion was once mooted by a prominent climber to build a concrete climb in Hyde Park. This suggests that many of the climbers regard a mountain as a mechanical exerciser. Esmé and I climbed often at one time. We were fortunate in numbering among our friends one of the most resourceful climbers in the country, who grounded us well in the principles of the sport.

The Welsh climbs vary greatly in height. There are difficult climbs of a hundred feet, and at least two climbs of nearly a thousand. Most of these climbs are classified and described in various handbooks, and are graded as 'moderate', 'difficult' or 'severe'. A climb is divided into 'pitches'. A pitch is the term given to the length of rock which must for convenience be climbed at a stretch. Suppose a party of three wished to tackle a six-hundred-foot climb whose longest pitch is ninety feet. They rope together, with a hundred-foot rope between first man and second, and another between second and third. Then the leader leaves the ground and climbs carefully upward. The first pitch, perhaps, is the long one of ninety feet, and is terminated by, let us say, a quartz ledge. The leader ensconces himself securely on the ledge, and if possible slips over a firm spur of rock the rope which reaches down to the next man. The second man commences to climb up to join the leader. And as he mounts, the leader steadily takes in the slack in the rope, hauling it over the rock bollard, if there is one, and possibly round his own shoulders as well. For a securely planted man can take a very strong downward tug without being pulled off. The leader in no way hauls up the second man, but he keeps the rope always tight, so that should his companion fall, he will do no worse than dangle like a spider. When the second man arrives at the ledge he will at once tie himself to a projection, if there is one, and the leader will commence to work upward once more. Assume that the next pitch is seventy feet. As the leader climbs, the second man sees that his rope runs free, for on delicate holds the least jerk is dangerous. And he sees, too, that the rope runs behind or over a projection. Thus, should the leader fall, he will in theory be brought up short by the rope. But if he has made only thirty feet of progress, he will have to drop for sixty feet before the rope tightens. The rope would probably break if the fall were direct, but perhaps the unfortunate man would bounce once or twice and dissipate some of the forces of gravity on the rocks. In this case the rope might hold. In any event the second man would not be pulled off if he had taken due precaution with the belay. But the leader rarely falls. He dare not afford the luxury. And when he reaches the end of the second pitch he belays himself while he waits for the man below him to bring the last one of the party up the first pitch. The last man reaches the first quartz ledge and ties himself on, to await the second man's ascent up to the leader. The good climber never takes a chance. There are a large number of irresponsible performers who think it either effeminate or troublesome to learn safe rope technique, but if these people are spared they receive sooner or later a fright which transforms them into careful climbers. And unless a rope is properly used it can be converted from a very real safeguard to a menace. The layman still seems to imagine that the rope is to ensure that the party shall hurtle to destruction as one man in the event of a slip by any member. No doubt this would be in the best traditions of British sport, but it is certainly carrying the team spirit to a dangerous extreme. Or else the layman believes that the rope is used to haul the dangling climbers up the rocks like sacks of coal. Presumably the Indian rope-trick is employed to aid the leading man.

When I first began to climb I had not yet reached the intersection of the graphs of age and sanity. I climbed at great speed and with a total absence of security. I had been climbing one day with an experienced friend on the six-hundred-foot rocks behind Glyder Fawr which are known as the Idwal Slabs. We had made several different ascents with my friend leading, and, being unconscious then of the infinitesimal margin between security and fatality, I remarked that I wished to lead myself for the first time, and up something more difficult at that. To the left of the Slabs is a very thin climb known as the Tennis Shoe. I believe it was first ascended by Odell, the Everest man. I proposed to go up it. My friend very rightly demurred.

'All right,' I said, 'you wait for me. I'll just go part-way up to have a look at it.'

I cast off my rope and began. The climb was well off the vertical fortunately. The main feature was a narrow crack which enabled one to enter the fingers of both hands and to maintain position by pulling sideways with each hand simultaneously. There were few footholds, and the feet could only assist by the friction of rubber soles on the rock. Worse, the crack began to fade away. The lichen-covered rock told me that the route was but rarely used. The beginner finds descent very difficult. I took one look between my legs at the inhospitable rock-face which fell away underneath me, and decided that to go on, though probably impossible for a tyro, was preferable. But help was at hand. A few yards to my right lay another, much easier climb. My companion had made his way up this until he was above me and to one side. He swung the rope down to me.

'Tie it on, just as a matter of form!' he called.

I managed to detach a hand from the shrinking crack, and to make a clumsy but very secure knot around my waist. I gained a few more feet of height. The crack vanished. I managed to get the fingers of my left hand over a small hold. Above it a series of holds led upward. It seemed to me that if I could but draw myself up a little with the left hand I might reach the first of the series with my right. But I dared not move strenuously; my purchase was all too small. And slowly my strength gave. The frantic fingers of my left hand straightened gradually. I gave a despairing cry to my companion, who seemed to think my tribulation entertaining, and off I fell. But I only swung across the rock on the stretching rope like a pendulum. I had learned my lesson cheaply. Esmé is a very good climber. She is strong for her light weight, and enjoys a perfect sense of balance, so that she prefers the delicate type of climb which calls for balance and smooth movement, while I am more at home on the strenuous pitches which demand muscular effort rather than finesse. But I do not think that climbing has ever had so strong an attraction for either of us as has mountain-walking. However, the sport is bound up in the history of our district, and we cannot be indifferent to it.

There are many climbing accidents, of course, but nearly all of them provide those involved with opportunities to show great courage and devotion. This in itself is sufficient answer to those who say that climbing is a pointless pastime, that it is not even the quickest way up a mountain. The human peoples must have a high-water mark of spiritual attainment at which to aim. Mallory set a standard during his rescue of the porters on Everest; so did Captain Oates on his last walk in the Antarctic. Examples such as these give spiritual exercise to all who hear of them. I know of only one Welsh climbing accident where the conduct of the survivors deserves criticism. But that criticism must be of the strongest.

On a rough November day in 1927, four men set out to climb the Great Gully of Craig-yr-Ysfa. The Great Gully is one of the longest climbs in the British Isles. It is difficult when dry, and extremely difficult when wet, as it was on that day. The leader of the party was a very fine climber named Giveen. He chose to take with him on this severe ascent in such bad weather three novices, Stott, Taylor and Tayleur. The party were staying at the Climbers' Hut at Helyg, and were no doubt quite fatigued, wet and cold before they reached the foot of the climb, which lay two hours' walk away over the Carnedds. The rain and wind so delayed the three inexperienced men on the actual ascent that it was seven o'clock when Giveen led them to the top by lantern-light. All four were soaked to the skin, half frozen with cold, tired out and faint with hunger. They began to scramble down off the Carnedds on their long walk back to the Hut for shelter, food and warmth. And then Tayleur dropped the compass! They could not find it, and struggled painfully on. Almost at once the lantern gave out, and in the darkness Stott and Taylor blundered into a lake. Stott managed to scramble out of the icy water, but he heard Taylor still struggling, and dived back to pull him out. Both men collapsed on the shore.

Previous to this disaster Tayleur had been the most distressed man of the party, for he was not physically so strong as the others. Giveen decided that Tayleur too would collapse unless hurried to shelter. He dragged Stott and Taylor behind a wind-break, where they would be to some extent out of the wind and rain, and then assisted the failing Tayleur towards the Hut. It was another four hours before they reached it. They stayed there only long enough to snatch a bite of food before getting out a car and driving to a hotel some five miles away which was, and still is, kept by a climbing man who has considerable experience of rescue work. A party was organised at once, but the croaking of ravens led the searchers to the frozen bodies of Taylor and Stott. That was the story which Giveen told later at the inquest.

Everybody was sympathetic except young Stott's father. He insisted that it was madness to take three novices up so severe a climb in such weather. And soon his criticisms became more pointed. Probably instinct told him that there was something amiss. He persuaded friends to visit the scene of the disaster, and to calm him they went. But they returned in a very different frame of mind, for on the trampled lake shore, stopped by immersion, lay Taylor's watch. The hands said 6.40. Through the Press old Stott challenged Giveen to explain what he had done during the twelve hours which had evidently elapsed between the accident and his arrival by car at the hotel to demand help. Ugly rumours became current locally. Tayleur, the other survivor, when questioned, began to admit that he had not been particularly exhausted that night, and that there had been no need for Giveen's solicitude. Finally he said that Giveen's statement was 'a damned lie'.

The rescuers then began to recall the finding of the bodies. The dead men had been lying face down in a bog, just as they must have collapsed. Their equipment was still on their backs. Taylor was smothered by the peaty mud. It began to be clear that Giveen, anxious for his own safety, had persuaded the weaker-willed Tayleur to hurry off with him, and had left the bodies where they had fallen. Stott's father now had stages of the route timed. Under the most adverse conditions the Hut was only a two-hour journey from the top of the climb. It was apparent that Giveen and his companion had reached shelter at about 9pm, and after a meal had slept through till morning. There are farms within ten minutes or so of the Climbers' Hut. Giveen's inhumanity was so callous as to be insane. Even when he had at last gone to the hotel to report the accident he had ordered breakfast, and had conveyed no impression of urgency.

Many members of the Climbers' Club threatened resignation unless Giveen were expelled. The Club held an inquiry, but before they could reach a verdict the defendant himself marched in and resigned. He alternately laughed at the members and cursed them before he left the building. So universally was the man condemned that later he was turned out of the house by the poor landlord of a remote mountain inn, who refused to serve him or to take his money. But there had been probably more than a glimmer of insanity in Giveen's mind that night. Not long afterwards he was a patient in a mental home, and on his release he met his death by suicide.

Of course, it is easy to be wise after the event. If Tayleur had really been physically distressed Giveen might have been justified in hurrying him off in order to save at least someone from the disaster. But even so he should have spared some of their clothes for the other two, and should have found help for them without the delay of a second. If one man of a party of two meets with an accident, the sound man is in a difficult position. He must decide whether to leave his companion and seek help, or whether to stay with him and trust that they will be missed and searched for. Quite recently a party of four people were descending Tryfan. Two were some way in front of the last pair, who were a young man and a girl. Mist came down. The girl fell. She fell only a few feet, but her head struck a rock. Her companion dragged her into a sheltered place and wrapped her warmly in some of his clothes. He waited to see whether their companions would return, but no one came, and dusk fell. The girl never became sufficiently conscious to under-

251

stand him, but at last he left to fetch aid. Later, when he guided back a party with lanterns, the girl had vanished. They searched all night, all the next morning, and only in the afternoon did they discover her dead body. She had risen groping to her feet in her delirium, and had fallen many hundreds of feet. There is no doubt that in this case the young man should have remained with his friend. The other two members of the party would inevitably have returned to look for them sooner or later. But one cannot blame the young man for his mistake. He must have felt that immediate aid was essential for his unconscious companion, and judgment easily becomes warped under such difficult conditions. I think that if I were left with an unconscious person without reasonable hope of quick aid I should tie the hands and feet, dress him in most of my clothes, and leave him in a comfortable and sheltered place. It would be almost an impossibility for the strongest man to carry down a body from the mountain.

The most difficult cases for injured and rescuers alike are those where the missing people have not left word where they are going. No real walker or climber will leave home without telling someone his plans for the day. It is selfish not to do so, and has often led to endless trouble. Sometimes it is difficult to adhere to these plans. Some years ago two men, both good climbers, left their hotel in the Quellyn valley, west of Snowdon, to do some climbs on that side of the range. Presently they found themselves on the Crib-y-Ddysgyl ridge, and dropped down towards the Llanberis Pass to do one short climb there. One of them fell. It was a difficult place. The second man was insecurely belayed, and could not hold the leader. He was pulled off too. One man was instantly killed. The other lay conscious but too badly injured to move, roped to his dead friend. When they did not return, search was made for them during the night on the Quellyn side of Snowdon, where they were known to have gone. The search was intensified at daybreak, with obvious lack of success. The survivor lay out in the rain for two nights and the best part of two days before an old woman came across him while looking for a strayed cow. He was still conscious, and he recovered, with no visible damage beyond a limp.

Even in the depths of winter news of an accident flashes through the district by some form of telepathy. Once in winter I was called to help carry an injured man off Tryfan. He was one of a party of three climbers, and had been struck on the head by a falling stone which had knocked him senseless. He had slipped at once from his holds, of course, and had nearly pulled the other two with him. They had had great difficulty in lowering him to the ground. I collected Thomas, who insisted on wearing rubber thigh-boots, and ran round in the car to Ogwen. I had been sent for because there were supposed to be no visitors in the district at that season, and because few local people had climbing experience. But the foot of Tryfan was like the car-park at a tattoo. Thomas and I met the rescue party when they were halfway down. There were at least thirty people present, and many more sightseers were waiting in the road. Thomas, slipping and tumbling among the rocks in his rubber boots, was nearly another casualty. Those who were not engaged with the stretcher were having an excellent time fortifying themselves with the brandy which everyone had had the foresight to carry up with him.

On yet another winter day at Dyffryn I was visited by an AA scout who reported that a motor-cycle belonging to an acquaintance of mine, one Macdonald, had been lying by the roadside under some cliffs in the Aberglaslyn Pass since the previous afternoon. I drove down there at once. The bicycle was soon found. Snow was deep on the hillsides, but most of it was fresh fallen, and all tracks were obliterated. I spent an hour or two searching around rocks up which Macdonald might have climbed. I came on no signs of tragedy, and presently returned to the bicycle. I looked it over for a message, but found none. I returned to Capel Curig, and called to tell the facts to the same hotel keeper who had fetched the bodies from near the Great Gully. He knew Macdonald, and told me that he spent much time scrambling about among the cliffs down there. Word was sent round to collect more searchers. Meanwhile the proprietor said, 'He lives with his people. I'd better phone them and prepare them for a shock.'

I heard one side of the conversation.

'Hullo! Can I speak to Mr Macdonald? That Mr Macdonald? I am speaking from Capel Curig. I am afraid your son has had a slight accident. What! He's painting in his studio! Oh! It's a mistake, then.'

Master Macdonald had left his motor-bicycle, which had refused to start, managed to get a lift to his home thirty miles away, and had told no one in Capel Curig.

But the percentage of accidents among climbers is remarkably low considering that the standard of climbing is being raised each year. For the most skilled men are today completing climbs which are on the limit of what is mechanically possible.

When I had first met Esmé I asked her to come to spend the next day with me at Capel Curig, and to pass part of it climbing. She duly arrived, and I drove her down into the Gwynant valley with the intention of ascending a climb known as Lockwood's Chimney. This is a very pleasant and quite easy scramble. The first pitches run up slabs which have plentiful holds. Halfway is the chimney itself. Up this, one proceeds in the time-honoured method employed by the sweeps' boys of Dickensian times. Presently the chimney turns well into the depths of the mountain, to lead at last back again to a high ledge on the face. And from the ledge one climbs the last pitch, an easy slab, to the sloping grass of the hillside above.

At the foot of the climb I politely explained to Esmé the first principles of rope technique. She listened just as politely. Both of us were very polite. I went up the first pitch and belayed. Esmé came up as fast as I could haul in the slack of the rope. I gave up my stance to her and climbed again to the foot of the chimney. She came up like a lift to join me. I wedged myself into the chimney and endeavoured, even at the expense of barked shins and elbows, to make an ultra-rapid ascent. I went so fast that I almost bumped my head on the chock-stone. The chock-stone is a boulder which in ages past tumbled down into the crack and became wedged. It is expedient to wriggle belly-down like a snake over the stone in order to commence the inward climb along the chimney.

'It's rather a sweat!' I called to Esmé. 'I'll give you a pull if you shout.'

But again the rope slackened in my hands, and Esmé's unruffled features and detached smile appeared above the chock-stone, which she surmounted with none of my clumsiness. I grunted, and negotiated the rest of the chimney until we stood on the ledge. The ledge is a grand place. From it, one looks down the valley over the beautiful tree-framed Gwynant Lake to Moel Hebog. The scenery is softer and more gracious in the vale of Gwynant. But on the other side one can see our own fierce hills, Siabod and the Glyders. Esmé sat on the ledge, which is two or three feet wide, and dangled her legs. The steep-climbing Gwynant Pass was almost level with us on the other side of the narrow valley. Hundreds of feet below our feet the young Glaslyn river hurried to refresh itself in Gwynant Lake before going to seek its fortune in the sea.

'You climb very well,' I remarked, not too pleased perhaps at this proficiency, which was fully equal to my own.

'I've done quite a bit,' answered Esmé. 'As a matter of fact, I've been to Capel Curig before. This climb was the first one I ever did.'

So even in those early days Esmé never gave away much gratuitous information.

21

The Annual Sale

FROM MARCH ONWARD THROUGH THE summer we have been busy at Dyffryn. We have fought many enemies – stoats, foxes, crows, snow, rain, floods, marauding dogs and the many difficulties of our rough, rocky land. We have felt and resolved many anxieties, known despair and relief, failure and triumph. Everything happens big at Dyffryn. And our biggest day and our most anxious day crowns our working year. In mid-September we hold on the farm our annual auction sale. On this one day, in the space of two or three hours, we learn what income we may expect for the next twelve months. This day is the culmination of all our endeavours. To its end we have worked through the lambing, the cutting, the washing, the shearing and the dipping. Last year's hay has been snatched from the weather to feed this year's sale cattle. Errors or successes in breeding meet their reward now. Economic or political changes bend all their influence for one day on our remote valley.

At the two dippings we have already picked out our four-year-old ewes and our wether lambs, and they are all grazing together in the ffridd. A couple of days before the sale our patient neighbours come to help. We gather the ffridd and drive the sheep into the pens. At once we sort out the wether lambs, and leave them in a side-pen out of our way. Then we turn the three or four hundred ewes all together into the largest pen. They are not too tight-packed there for us to have difficulty in viewing them individually, and not too slack to be difficult to catch. Every Welshman is, self-appointed, a judge of sheep from whose final decision there is no appeal. Tact beyond the dreams of diplomacy is necessary throughout the day.

'The best pen first,' cries Bob Henblas, lighting his pipe with the relish of anticipation. We sell our sheep in lots of twenty, and it is most important to match the individuals of each lot exactly. Even in a carefully bred flock there is to the trained eye great variation between ewe and ewe. Buyers have preferences for certain types, and bid well for their fancy. But a lot made up of mixed types would have small appeal.

'All right,' I say, 'we'll pick the biggest and heaviest.'

Year after year the procedure has been the same. Each man plunges into the mêlée and picks out a ewe on which he has had his eye. We turn these proud matrons into a smaller pen, until about twenty-five have been collected.

'Five must come back!' says John Davies.

There is always more dispute over the removal of the five than there was over the selection of the twenty-five.

'That thing like a camel must come out!' I insist.

'By damn!' exclaims Thomas. 'I have pick her. There ben't nothing like her.'

'There isn't!' I agree. 'She's as tall as the rest, but she's all legs and no weight.'

But in the end by statecraft and by cunning a fine level pen is made ready. These twenty sheep are the pick out of the mature ewes of a very large flock. They bear inspection, and it is hard to urge the men back to work.

'Now,' I say to Thomas, 'we'll have your tall leggy ones.'

And while we continue to pick, two men mark the first pen. I always use coloured paint to mark. We employ four colours. In order of merit they are red, blue, green and yellow. The first pen is marked with a dot of red on the top of the head; the next on the shoulder; the third on the middle of the back; the fourth on the rump; the fifth on the tail itself. Thus each colour deals with five pens, or a hundred sheep, and our four colours clearly distinguish four hundred animals for us. Some farmers stamp each pen with a number. But if perhaps a purchaser wishes to resell his pen quite soon, the number is a disadvantage to him. The crowd will say, 'Ah, these are number eight from Dyffryn sale! The first pen' – whose price will be known to the whole countryside – 'fetched so and so, therefore the eighth will have cost about such and such.'

No vendor likes buyers to have a price limit at which to aim. So our more anonymous form of marking finds great favour, and is perfectly clear.

When the second pen of big leggy ewes is matched, it looks very nice. There are no heavy types to form a contrast which would give the others a weedy appearance. But the third pen is the choice of the day. Many among the first lot of fine, strapping ewes have a good reason for their physical excellence. It is more than likely that they have been barren for some of their breeding years. They have escaped the burden of gestation, and later of suckling. Thus their development has not been retarded. And the next pen are not ewes of a type which we hill people like. But they are too tall to be ignored, and must take second place. However, the third pen is made up of ewes of compact body and good wool. These sheep are nearly as large as the first pen, and are all good mothers. They are nearly as strong as the best pen, in spite of the burden of mother-hood. The buyers, who are no fools, know this, and often the third pen equals or beats the other two in price.

When three or four pens have been carefully graded, we pick out thirty or forty of the poorest sheep from the balance of the flock, and pen them temporarily out of our way. With the best and the worst taken from their midst, the residue are very level. We drive them without any selection in twenty-fives into a side-pen, and turn back each time the five who are most out of sympathy with the general appearance. The work now can proceed at a great pace, and speedily we find that the markers are using the yellow paint. With my system of carefully culling out unpromising young sheep there is usually a pen or two of these. We mark these differently, maybe with stripes. For their youth is a great selling-point, and when they come to be turned into the sale ring I shall wish to announce what they are.

Some years I have a quantity of sheep to sell which I have bought to graze the summer grass on the enclosures below the mountain. We mark these sheep in the same way, for their various earmarks differentiate them from the Dyffryn sheep when we come to sort them out on sale day.

After the ewes come the wether lambs. There is less difficulty here in the choosing. It is almost altogether a question of size now, and pen after pen is speedily selected and marked. When they are done we fetch up the ewe lambs from the bottom lands where they are grazing while they await their trip to the wintering grounds. I cull through them carefully, and pick out any that do not reach the highest standards. The more conservative of my assistants protest against this sacrifice of youth, but results have long since proved my case, and I carry on. These again form a choice lot. Well-bred ewe lambs are hard to come by, and the lowland man will pay well for my rejects. For, while these culls might die under the rigours of the Dyffryn climate, they will grow and thrive in the milder conditions lower down.

And then we come to the rams. This is the most controversial operation of the day – and that is saying much. Here everyone must have his say. Davies drives more than forty haughty males into the pen. Big of bone and body, and with widespread twirl of horns, the rams fill the sorting pen. The tupping season is only a few weeks away, and they are touchy and quick to take offence. Pairs of them sullenly batter each other as we watch, until we interfere with their fights. Since we do not have our ewe lambs served we can keep rams for two seasons at Dyffryn before they have a chance to inbreed. The best of them we keep for three seasons, for a little inbreeding where the most desired type is employed tends to stamp on the flock the qualities which one wants. Most of the rams have been bought. They come from various other mountain flocks which have impressed me by their good qualities, and they all have desirable heredity to transmit. But a few rams are Dyffryn-bred. We do not turn these among the flock which will in early November be gathered into the ffridd for tupping, but we send them up on the mountain to serve those ewes which the gathering inevitably misses. Obviously, bought rams would, as strangers, be useless for this purpose.

'That old warrior have been here three year,' says Davies. 'He come from Dinas Mawddwy that time. Let the flamer go.'

'All right,' I agree.

'And that polly without horns – it ben't worth potching with him. He won't stay home. I have tramp miles after the flamer.'

'He's a grand ram,' protests Gwylim Pantffynnon.

'He do suit you, I do know that!' answers Davies. 'He have been on your place when I lost him last season and have got your best lambs for you.'

'Buy him, Gwylim,' I suggest.

'I must ask the boss,' says Gwylim, discomfited.

Presently a dozen rams, for reasons of age or unsuitability, are waiting for the mark. Amid a babel of protest or acclamation I direct the order in which they are to be marked. Then the whole multi-coloured flock is turned back into the ffridd to await the great day.

The sale is extensively advertised in local newspapers, posters have been pasted on hoardings, and leaflets distributed at marts. In addition several hundred cards are printed. A copy of one is given below.

BY DIRECTION OF MESSRS. E. & T. J. FIRBANK.

Dyffryn Mymbyr, Capel Curig

ANNUAL SALE

1,050 GRAND HILL SHEEP

INCLUDING

450 STRONG HEALTHY EWES
125 YEARLING WETHERS
50 CHOICE EWE LAMBS
400 FORWARD WETHER LAMBS
12 STOCK RAMS

ALSO

21 ATTESTED WELSH AND CROSSBRED CATTLE

COMPRISING

2 Calving Cows

17 Two years and 1½ years old Bullocks and Heifers
WHITE SHORTHORN STOCK BULL
WELSH BULL CALF, 10 months old

USEFUL HALF-LEG GELDING, 16 hands high
(Good worker in all gears).

TO BE SOLD BY PUBLIC AUCTION AT

DYFFRYN MYMBYR

AS ABOVE BY

ROBERTS & ROGERS JONES, F.A.I.

ON

SATURDAY, SEPTEMBER 13th, 1941

AT 12-30 O'CLOCK P.M.

USUAL CREDIT TERMS. REFRESHMENTS PROVIDED.
Free keep will be allowed to purchasers on all stock sold, if required.
Every facility will be given for the delivery of stock to purchasers.
Crosville Bus will leave Llanrwst Square at 11-20 a.m. for the place of Sale.
Auctioneers' Office: Ty'n-y-Fynwent, Llanrwst.   Telephone No. Llanrwst 18.

J. Lloyd Roberts & Co., Printers, Llanrwst.

I keep a list of the addresses of all the men with whom I have had dealings from time to time, and who may be interested in my sale. Esmé and I address and dispatch about two hundred of these cards, and the auctioneers send out many more. Our sale is a credit sale. Because of the impoverished condition of British agriculture, credit terms are a magnet. Prices are higher than at a cash sale, it is true, but by the time that the debt falls due the ewes bought have lambed, the wether lambs have fattened, the cows have calved, the bullocks are beef. So that the money should be ready in the bank for the buyer to meet his obligations. There is no risk for the vendor in these credit sales, which seem peculiar to parts of north Wales. The auctioneer is responsible for paying over the total sum due, less selling charges, on the appointed date. And it is entirely the auctioneer's business to see that he is recouped by the various purchasers. Every old established firm of auctioneers has its regular following of buyers. Difficulties are so great in British farming that there has to be much give and take between the different sections of the industry, and by long association the auctioneer knows who is trustworthy and who unreliable. Thus credit is often allowed to a man whose concrete assets are nil, but who has the more important asset of integrity. At these credit sales a small discount is usually offered for cash, but this concession is rarely more than two-and-a-half per cent. Ready money is worth more than six-pence in the pound interest to any farmer, so that only a very small proportion of the total sum involved at a credit sale is paid up at once. Most of these sales have a period of six months, a few have three months.

Esmé supplies free lunches and teas on sale day. The catering is difficult, for if the weather is bad only a handful of determined buyers will be present; if it is good, then two hundred people will be there, some to buy, most to watch. But we welcome even the parasites. For one thing, a farmer must attend sales to keep himself abreast of prices and to reinforce his judgment of beasts. And, again, few farmers can resist throwing in a bid now and then when stock is in the ring. On a pen of twenty sheep each shilling rise means a sovereign. I do not like a small crowd at my sale, even though there may be just as many genuine buyers present as on the fine days when the countryside treats Dyffryn as a rendezvous. For among a small number of people it is not difficult to discover who is bidding for some particular lot. Among regular sale-goers are many cliques and friends. When old So-and-so, who is a popular man, is seen to be bidding, many people refrain from raising him. But in a large crowd old So-and-so is more anonymous, and his dearest friend may run him up without knowing it.

The lunch which Esmé provides is quite sumptuous. It has, of course, to be cold, for even in the old farmhouse the visitors have to sit in relays. There is usually a large round of beef, though in some years we have killed our own mutton with success. There are elaborate salads, fruit jellies, milk puddings, custards, jam tarts, stacked loaves of bread for those who want to eat cheese and, inevitably, enough tea to float a battleship. After the sale, which Esmé has never yet found time to see, the mob flocks back to the old house again for bread and butter, bara brith – the currant bread – scones, jam tarts and more strong sweet tea.

I entertain the auctioneers to both meals in the dining-room of Dyffryn house. Lunch is a strained ceremony. Cassandra herself would not have deepened the portentous foreboding. The auctioneers talk of anything but agriculture. I talk of nothing. I envy Esmé her busy-ness, as I fidget and fiddle, and do sums in my head, and watch the weather, and count the arrivals, and wonder if the marked sheep have escaped from the ffridd.

After an era has passed, conclusively proving that time is relative, I find myself with the auctioneers at the pens. Cars and special buses are parked along the main road some way beneath the pens. A thin stream of farmers trickles down the rough hillside from the old house. The advertised time of the start is never adhered to, and not until a few hastening stragglers reveal that the dregs of the crowd are on the way does the chief auctioneer clear his throat with a clarion note.

Meanwhile our usual gang of neighbourly helpers, unsuitably

dressed in their best suits, which are protected by overalls, have sorted out the first six or seven lots of sheep. We sell in a wire-netting pen constructed outside the stone pens. The first fine twenty ewes, bearing the red mark on their heads like a decoration, are the centre of all eyes. Our careful matching stands them in good stead. As they group and regroup and circle about, there is no telling the one from the other. The only standard of comparison is excellence. The sorters can be idle for a moment before the start, for there is no further accommodation in the main pens for more separated lots. But when the selling begins the men will have to work like demons to sort quickly enough to feed the Moloch ring, for it is bad to have a check at a sale. There is an electric atmosphere generated at an auction which must not be dissipated. The long climax of bid, counter-bid and bang of the gavel must be sustained from first to last. What connexion is there between a farm sale, mob hysteria, religious ecstasy, yoga, the immunity of fakirs, hypnosis? There is a connexion.

And presently the chief auctioneer will stroll into the ring.

Always he announces first in Welsh and then in English the conditions of sale, and explains the credit terms, lest some stranger should be there, flown with ignorance and gold. I have had eight sales now, two alone and six with Esmé. The procedure has to me become stylised. That short afternoon spent annually beside the pens holds always the culmination of endeavour, the resolution of anxiety or the despair of hope. The patter of the auctioneer, the chatter of the crowd, the very actions of the startled, patient sheep in the ring, are typical and unvaried. The intensity of the happening yearly impresses itself on my consciousness, just as the seasons are tallied by rings in the trunk of a tree.

'Now!' says the auctioneer, when the preliminaries are done according to rote. 'Now! A fancy pen! Mr Firbank guarantees all sheep sound in eye, tooth and foot. If you find a wrong 'un you can have her for nothing! Now what'll you bid?'

There is no enthusiastic outburst. The first pen sets the standard for the sale. The buyers mentally nudge one another. None comes forward. But my guarantee always inspires confidence. At a large sale, especially where the flock is not penned in lots for inspection, it is very difficult carefully to look at the mouths of the sheep for age, at the shade of the white of the eyes for health, at the hoofs for signs of foot-rot. There might well be diffidence to restrain full-blooded bidding but for the guarantee. And only once have I been called on to honour it. A sheep sold as sound had a 'broken mouth'. This is the term for lost teeth. On upland grazing sheep keep their teeth to a very ripe age, but now and again a young ewe will lose some through injury.

'Come on!' calls the auctioneer. 'Put them in at your own price then! Right off the mountain! These sheep have wintered at home since their first year! They'll improve every day! What shall I say, gentlemen? Thirty-two? Thirty-two, yes? All right, twenty-six! Twenty-six shillings I am bid! Seven! Eight! Nine! Twenty-nine shillings! Bid up! We've a lot of work to get through! I'll take your sixpences now! And six! Twenty-nine-and-six! Level it up! Thirty! And six! One! Thirty-one! Don't miss these! The pick of the flock! And six! Thirty-one six! All right then! Nine! Was that a bid, sir! Thank you! Thirty-two! And three! A fresh bidder now! And six! Thirty-two-and-six!'

The auctioneer pauses a moment. The bidding is obviously close to its climax. He leans forward confidentially.

'Look at their type, gentlemen! Look at their breeding! Thank you, sir. Thirty-two-and-nine! It's against you now, sir! Don't lose them for a few pence!'

With a gesture of finality a man nods a last bid, and immediately turns his back, as if he were going to take no further interest.

'Thirty-three shillings! I am selling at thirty-three shillings!

They are going at thirty-three shillings. Are you all done at thirty-three?'

Smack! The auctioneer's stick comes down on his leather leggings. Twenty sheep are sold at thirty-three. I have gained thirty-three pounds cash and lost one or two avoirdupois. Sometimes at sales the price of the first pen is fictitious. Vendors know well how important it is to set a high mark at the start. Often they have a regular customer who has bought for years their best lots. Privately they offer him a pound, or even two, luck money if he buys. Thus the customer can bid up to a shilling or two over the market price without loss to himself, and the sale may well proceed at an artificial level. Each year our average price alters. At my first sale in the slump year of 1932 my best pen fetched little more than twenty shillings, and in 1937 the same class of ewe sold for over forty. Economically I feel that the Welsh mountain ewe is worth about thirty shillings, and the wether lamb about nineteen. At these prices both types should pay the mountain producer, the lowland farmer, and later the butcher, who will be able to sell the meat to the public at a fair price. Our best hill wether lambs are almost ready for the butcher in September, the poorest may take till spring to fatten. The ewes are bred from for one, two or more seasons before they are fed up for mutton. The sale goes on. For me, much of the tension is over when that first pen has passed to new ownership. I know what I can expect for all the thousand or more sheep which remain. Perhaps the second or the third pen sell for equal money to, or even more money than, the first. Then the price grades downward. There may be a fall of sixpence or a shilling on each lot. Then for some reason a pen will catch the fancy of two or three buyers, or it may be that the competitive spirit develops between some obstinate men. The price of the lot rises above that of its immediate few predecessors. At last the final pen of full-aged ewes is passed through. Amid some hilarity a frugal purchaser winks the winning threepence. Among this lot, no doubt, is a ruptured ewe, a ewe whose broken leg has reset crooked, unseen by us, a ewe whose udder is not normal, or whose teeth are announced as broken. And then come the pen or two of young ewes.

'Now here you are, gentlemen!' says the auctioneer. 'Nearly all yearlings and two-year-olds. You can't go wrong. Fine, well-bred stock! Just the thing for the lowlands! Mr Firbank's only selling them because the wool is a bit soft for the mountain! Shall I say thirty shillings? Thirty, yes? Very well, twenty-six – it's for you to say! And six! Twenty-seven! And six! Twenty-seven-and-six, genuine bid for these young ewes! Threepence! Very well! Twenty-seven nine! Twenty-eight! Come along now! You can't buy young sheep every day! These will – two bids of threepence! Can't accept that! Which of you gentlemen will say six? Thank you! Twenty-eight-and-six! At twenty-eight shillings and six-pence they are going! They are going at twenty-eight shillings and sixpence! I am selling them at twenty-eight-and-six! Are you all done?'

Smack!

And if there are any ewes which I have bought for the summer grass, they come next. Sometimes their previous owners are present. If I make a good profit, they tell me I must buy them a drink next time we meet at the mart. If my speculation has been poor, they commiserate with me. The ewes finish. The first pen of wether lambs comes leaping into the ring. The crowd, in the irresistible way which sale crowds possess, has encroached on the ring by now. Men line the inside of the wire-netting. The bouncing lambs hustle behind the spectators. They leap uncontrollably, to cannon into important stomachs or to flash between bow legs. They have no cohesion, and are coaxed to stand in the centre of the arena only with difficulty.

'Well!' says the auctioneer. 'I've never set eyes on a more promising bunch! These lambs are fat, gentlemen! They'd cut up today! I can't start them below twenty-five!'

A buyer, bold to show his exact judgment, nods. It is usually a butcher who buys these best pens. It is an advertisement for him if it can be said that he has had the best pen at such and such a sale.

'Mr Jones's meat is always so good,' the housewives tell one another. 'He won't buy just anything.'

'Twenty-five!' calls the auctioneer. 'And six! Seven! Twenty-seven! Threepence. Twenty-seven shillings and threepence! You won't be sorry, sir, when you see these at home! Is there three-pence anywhere? I am bid twenty-seven-and-three for this beautiful pen! Fit to go anywhere! They're meat today!'

But the bidding was begun quite close to the mark. There is no addition from anyone. The stick cracks against the leather legging. The pen is turned, leaping, from the ring. More wether lambs are turned in, inspected, bid for, sold, and allowed to escape. Halfway through their sequence their value falls to the sovereign mark. By threepences, or sixpences, or shillings it lowers pen by pen to fifteen shillings or so. The last lot is turned in. They are always small lambs. Most of them have been born late, maybe at washing time, and they are young. One or two will not be quite sound, with minor deformities such as a short lower jaw or bandy legs. But such scrap lots have an irresistible appeal to certain men, who feel that low price necessarily means a bargain. This is sometimes so, it is true. But the only certain bargain is the best quality.

The pen or two of ewe lambs comes next. These are rejects from the Dyffryn flock, but they find ready purchasers, because ewe lambs are scarce in the market. And with the mild lowland climate and nutritious grass to build their physique they will produce many crops of lambs.

And at the end is the sale of the rams. John Davies comes into his own now. He always appears personally with his charges. Number one heaves into the ring. Davies, holding tight to his horn, is towed after him, and the ram is posed in a dozen tableaux before the critical audience. Someone, unasked, says, 'Two pun!' John Davies casts an incredulous look at the bidder, and immediately afterwards ignores him. He shows the ram's sound teeth to every one, he drags the beast close to the crowd so that they can feel and admire the hard, close wool. He expatiates on the fine length of tail, on the handsome, yellow-flecked face, on the sturdy leg-bones, the deep body, the arrogant Roman nose, the bold eye, the black nostrils, the clean hoofs and the broad forehead. The auctioneer, relegated to playing second figure, keeps muttering the last bid to himself.

'Fifty!' cries someone. John Davies and his charge at once besiege the man. Each succeeding bidder has to listen to a reiteration of the ram's excellence, until at last some startled farmer finds himself the owner at five pounds. And so through the dozen or more beasts.

Meanwhile Thomas and some assistants have gone to fetch the cattle. Usually there are several cows who are on the point of calving. Throughout the summer I am on the look-out for September calves. Then there are fifteen or twenty Welsh Black bullocks, who have been cropping the tussocky grass below the mountain wall, and there are several Welsh Black heifers, who are just about ready for the bull. We sell the in-calvers first, singly.

I pull out my notebook and check the dates for the auctioneer. 'Grand roan second calver!' he calls out. 'Only a week off her job! She'll do a fine milker, gentlemen! Look at her bag! Now, how much shall I say for this young cow? Twenty-five, yes? Well, then, twenty-four? Come along now! Put her in where you like. Twenty? Thank you! Twenty! Twenty pounds I am bid for this! One! Two!

Twenty-two pounds for this grand type of beast! I've never seen a better bag! Ten shillings! Very well, sir, ten shillings. Twenty-two pun ten! Twenty-three. Five! Twenty-three pun five!

Three of you in now! Was that a bid? No? Twenty-three pounds five shillings!'

And so it goes on. Every gesture, inflection, bid and counter-bid has become unforgettable to me. The only yearly variation is the price level. But that, unfortunately, is the most important thing. The bullocks go. The heifers go. The sale ends. In comparison with other auctions we always have a good sale at Dyffryn. Neighbours crowd round to offer congratulations. The bulk of the people drift up the hill to the old farmhouse for tea. It is still a long time before the auctioneers finish their last lingering cup with me in the dining-room, and before Esmé can leave her diminishing stacks of loaves and scones. We sit down eagerly then with pencil and paper, and quickly learn what our financial position will be during the next twelvemonth. Triumph needs celebration, disaster needs oblivion, so whatever the result we make merry that night.

And next day the men are early at the pens. One lot of sheep and cattle is driven off towards Llanberis, where the buyers from the Caernarvon district will meet it to collect their purchases; another lot goes past Ogwen to Bethesda, where some are put on rail for Anglesey, and others are met by more local farmers. But the bulk of the sheep and the beasts go bobbing down to Llanrwst. They are driven into the mart and separated into their various lots. A clerk of the auctioneer's checks out each pen and each beast to its lawful owner, and presently our responsibility is ended. It is time to begin working for the next sale.

22

The Rams

OUR SALE SHEEP ARE GONE FROM DYFFRYN in mid-September, and there are, too, fewer cattle to watch. At the beginning of October our ewe lambs troop off to Llanrwst to be collected by the various lowland men who are going to winter them. For good or ill the weight of sale day is off our minds, and we can turn unharassed to the most pleasing routine work of the year. For it is time to buy in new rams. There is an atavistic satisfaction in ensuring the fecundity of female stock. It is an old and a true saying that the ram is half the flock, and the bull half the herd. At Dyffryn I was lucky to learn that lesson early, and quickly began to buy the best rams I could find. A splendid type of Welsh mountain ram can be bought for four or five pounds, while a mediocre animal costs two or three. A ram will serve at least fifty ewes, so that the difference in price between the poor and the good amounts to only a few pence per ewe served. Yet that ewe's lamb is worth many shillings more if begotten by a first-class father. And in permanent flocks where the ewe lambs are kept to breed for three generations the heredity is widely dispersed in a short time.

It does not do to keep many rams of one's own breeding. A few help to stamp the type, but too many lead to serious inbreeding, with all its attendant ills. Good heredity is more important than good looks in man or beast, though if possible it is best to combine the two. But I would rather use a moderate ram out of a well-bred established flock than a Rudolph Valentine of mysterious antecedents. Improvement in the type of ram employed has worked wonders at Dyffryn. The sheep have increased in weight, there is far less trouble at lambing-time owing to the better stamina of the ewes, and the flock is hardy enough to winter at home.

There is a type of ram known as the improved Welsh, which produces weightier progeny and begets a finer type of wool. But this fine wool debars us from using the ram. Mountain farmers have tried the improved Welsh ram, and have found that their flock lost in hardiness and became unable to weather the severe winters. Many ewes had to be sent away to winter. So we have to be most careful in choosing rams to introduce into the Dyffryn flock. We first seek the good heredity of a proved upland flock, then the close fleece, then the sturdy body. There are not so many hill flocks from which to choose. In our immediate district there are fine sheep-breeders, but it does not do to buy from the same parts for too long. Fresh strains of blood are wanted. We buy a few good rams each year at local sales, but for the bulk of our requirements we go forty or fifty miles away to another Welsh mountain sheep district. This lies about the hills surrounding Cader Idris – Arthur's Seat – and Arran Mawddwy. It is a remote, unchanged part of Wales, full of bloody legend. We all love the trip to the lonely farms, where we are greeted each year as if our last visit had been yesterday. Sometimes Esmé goes with me, sometimes John Davies or Thomas. I usually tow a trailer behind the car into which I can pack six or eight rams. This limitation of space entails two visits if I have been unable to obtain many suitable rams nearer home. On hill farms the rams are gathered by October, and are securely fenced in so that they cannot prematurely serve the ewe flock. We do not want early lambs up in the mountains, for it is inhospitable to usher them too early into our hard world. April is time enough for the newcomers to appear, for there is more chance then that kindly spring will have appeared, to touch the budding lives with her wand. So we do not loose the ram until the first week of November. Thus in October the restless males are shut in from temptation, and they are to be seen on every farm.

It is not always easy to buy. That ram is young and is wanted for another season. This one is promised to someone else. The other was strayed all last autumn, and can remain at home another year. Then many are not of the type one likes. There is little bargaining in ram-buying. If an animal is good, both parties know the fact. The one does not care whether he sells or not, the other knows that he must buy his fancy. There are sonorous names in those parts, and lilting names too – Rhiw Ogo, Rugog, Maes-y-Pandy. One becomes so weighted with tea and paradoxical light-cakes that the climb to each new farm and each new meal becomes a problem which is literally insurmountable.

I was first introduced to this district by a neighbourly local farmer, a careful sheep-breeder, who generously taught me much about rams. But for several years he hinted to me of a Mecca to which few had penetrated, and of which the excellent ram district which he had shown me was as the poor husk is to the rich ear. As our friendship ripened, the veil was slowly drawn aside. There was, I began to learn, a vast farm buried amid high mountains. A patriarch ruled the steep lands with benevolent severity. In this uncharted fastness were bred rams before whose beauty the very stars dimmed their envious light. A long probation had to be served before the disciple might so much as make simple pilgrimage to the guru of the hills. The decision of the master was finality itself, and those cast out might never more aspire to own a ram from the fabled flock. Even the fortunates who found favour would perhaps have to wait some years before it was indicated that a choice of just one might be made from certain specified rams. There was no bargaining. The ram was worth so much. The pilgrim was privileged to pay at once without demur. These rumours filtered through the superstitious silence, until I was tantalised beyond endurance. And at last my friendly neighbour accosted me one day with serious mien.

'I have talked to Iorwerdd Pridderch about you,' he whispered.

'If you like to go next week with my son Madoc he will see you.' The Revelation was at hand, but the days passed all too slowly. The gods of the weather had evidently been propitiated, for we drove south towards Dolgelley and Cader Idris in warm sunshine, which is a stranger to our blustering October. We plunged into the maze of steep valleys behind Dinas Mawddwy.

Dinas Mawddwy was the haunt of the notorious red-headed banditti. The descendants of these not remote ancestors are still living in clannish seclusion. Dinas is a queer place, and queer events still happen there. I have camped in its dark, eerie hollow, and the night speaks. But we turned at last up a mounting lane. The lane petered out, and we abandoned the car on the green hillside. The contours of the country were peculiar. The hills were high, but more like rolling downland than mountains. The land seemed open, but visibility was in reality close. Madoc led unerringly over hillocks and across hollows, but always upward. The lower country revealed itself to us as we rose, and the road by which we had come seemed even more secretive. At long last Madoc and I came over a grassy rise within sight of a long, low-built farmhouse, which should by the eminence of its position have been visible earlier, but which only now came into our view because of some mysterious quality of the place. Farm buildings buttressed the house like chicks clustering about a hen.

We approached an open door and peered from the golden sunlight into the repose of a wide hall floored with blue-stone flags. The ticking of a polished oak grandfather clock welcomed us.

Madoc's quiet knock thundered, and at once a little, black-garbed, white-haired lady clattered over the stone floor to greet us. She spoke in Welsh, and bade Madoc apologise to me for her total lack of English. The master was out on the hill. He expected us. Would we go seek him? We left the house and wended on upward until, when I looked back, the buildings were gone, and nothing was to be seen but the crests of hills which gave promise of mysterious valleys. A few sheep grazed about us as we hurried on. For a while they were the only life we saw, then a dog appeared over a rise and trotted amicably towards us. He was a foxhound, and was quite incongruous in that setting. But more dogs heaved over the hill. There were an airedale, three brown Welsh terriers, two white fox-terriers and five sheepdogs. None of the sheepdogs was related to the others. They were comprised of an old English bobtail, a blue-coated, wall-eyed Shetland, a black-and-white Welsh sheepdog, a big brown-and-black sheepdog and a short-tailed little Pembroke Corgi, who was by nature a cattledog. Following the dogs trotted a mountain pony. He carried on his back a little old man. The man made to us and stopped. When he halted so did the dogs. The little man was dressed in well-cut breeches which ended in shining leather leggings. He wore a coat of hard Welsh tweed. His linen was spotless.

'How are you, Madoc?' he asked. He never spoke Welsh to us the whole day. Madoc greeted him and introduced me. We chatted among the three of us as we went towards a wired enclosure. Inside, the old man spoke a few words to a dog, who nodded comprehension and ran off out of sight behind a fold in the ground. The rest of the pack sat expectant on their haunches. The old man continued to talk to us. His conversation was mostly about agricultural conditions and politics. He was not only well informed, but was able to draw logical deductions from his facts. We were interrupted by the return of the dog. He had in front of him a battalion of rams, a brigade of lanced cavalry at the charge, a moving forest of heads and horns which undulated like tree-tops in a strong wind. There must have been a hundred rams there, as they slowed suspiciously and finally slid to a halt when the other dogs encircled them. For irrespective of breed the dogs all seemed able to work sheep.

I have never seen such rams. In travels up and down the country one would see a few fit to compare with them, but here were a hundred together. All were perfectly matched. That they were bred of the same ancestry from the one age-old flock was obvious. All of them were sturdy of body, big of bone, proud of head, and their wool bristled with the wiry white kemp which renders a fleece impervious to weather. The young rams had short, snow-white beards of kemp, which merged into the wool of their chests like the lace cravats of Regency gallants. And the old warriors had manes which ruffed out about their necks, and ran back in a fierce crest along their spines. Abraham himself could not have viewed the sight calmly. Madoc and I were openly excited. Our host sat his pony quite impassively, and gave us the history of ram after ram with the assurance of an Arab discoursing upon the pedigree of a stallion. He had promised two particular rams a year ago to Madoc's father. He showed them to Madoc. A boy was summoned by white magic, and dragged the two rams from the press. Aided by a dog, he set off with them towards the house. The old man would sell nothing else to Madoc that day, but dismissed the flock abruptly, and led us back to the house for lunch.

We were given luncheon in some style in a delightful oak-furnished dining room. The windows, cut through immensely thick, irregular walls, gave views of the old man's billowing domain. The atmosphere of the place was kindlier, softer than that of Dyffryn. The old man and his wife were mellowed to a slow serenity by the peaceful aura. Dyffryn is never peaceful. Even the idyllic warmth of a windless summer day is but a brooding stillness, as if the spirits of the place were whispering further mischief. Life at Dyffryn is an endless fight against most of the weapons which nature possesses. The constant hammer-blows temper the steel of one's spirit or break it. There is no middle path at Dyffryn, and no respite to lick wounds. But here I saw success, the crowning mercy of two lives which had been left to evolve unhurried by catastrophe. My allegiance to Dyffryn did not really waver. 'Escape me never!' says Dyffryn.

We ate cold roast mutton with baked potatoes and homemade red-currant jelly. It was not the mutton of the lowlands, but a meat so delicious and so tender that its delicate fibres dissolved in one's mouth. The old lady kept smiling pleasantly, and since she could not speak English did not converse at all. Her husband and Madoc and I talked of sheep and agriculture and politics. There was no wireless in the house, and I do not suppose that the old couple saw any paper beyond the local Welsh weekly, but the farmer was informed up to the minute, and had that unerring judgment which is the countryman's birthright, sired by Observation out of Patience. What a pity it is that the material for Cabinets must be wooden. Statesmén, not block-heads, could be sought out in the country. We finished our meal, and I for one was sorry. I could have talked to the old man for hours. But something was on his mind. He came to the point quite suddenly. 'Are you in a hurry?' he asked. We had both meant to visit other farms, but we said no. 'I will show you something,' said our host. Madoc's eyes grew wild with excitement. 'His special rams!' he whispered to me. 'He has a few hidden away, which no one ever sees!'

The pony was brought to the door. The old man mounted, and the dogs formed a bodyguard about him. Madoc and I tramped beside the cortege for about an hour. The light, turfy plateau rose and fell so perplexingly that navigation was as difficult as among the dunes of a desert. But here there was fertility everywhere – green sweet grass, a few late flowers, transparent streams wending with provoking mystery through the maze of hillocks, and all about were grazing contented, strong sheep. And in the end we came above a sunlit hollow which was fenced in by wire-netting. It was a saucer a quarter of a mile across. At its bottom was a rare group of oaks, growing high up for hardwoods, and through their branches came the glimmer of a pool. We made our way down to the trees. The old man spoke to two of the sheepdogs, who streaked off eagerly to right and to left. Quickly they came back, and with them were a dozen rams. There is no praise beyond perfection. The rams were superlative. We caught them one by one and handled them. Our genuine enthusiasm delighted the calm old man who owned them.

'Which is your fancy?' he asked me. I pointed to a two-year-old, not yet fully developed, but of fine promise. 'And yours, Madoc?'

Madoc also indicated a two-year-old.

'Come together next year, and take your rams then.'

At last the slanting sun warned us of the time, and we walked back in the stillness of the late afternoon. Not a soul, not a house, was to be seen from this high aerie, for the distant valleys plunged so steeply that the life which sheltered along their floors was hidden from us. We approached the house, and I was surprised to see that the main ram flock was penned in a yard.

'It you want a ram to take home with you,' the farmer said to me, 'you can pick one.'

He indicated about ten which were for sale. There were others there which I would have preferred, but the ones among which my choice had to lie were only less attractive by comparison. In ordinary company they would have been superb. I picked my ram.

'You can pay me three pounds,' said his owner. And this was certainly cheap, for I would have given much more to introduce this strain into the Dyffryn flock. It was now time for food again. We entered the darkening hallway for a cosy meal by firelight in a parlour-kitchen. The room was typical of many such, but the dresser was more highly polished, the pewter on it was heavier, the grandfather clock was more dignified, the willow-pattern tea service more ancient, than any others I had seen. It was a pity that we had to hurry. I had little time in which to probe the secret of the quiescence of the place and of these people. But I shall go back.

In the darkening shadows Madoc and I stumbled away over the hills, each of us hitched by a rope to our straining rams, and the spell which had been laid on us was not broken until, hours later, we had reached home.

The rams, old and new, are penned in a high-walled enclosure at Dyffryn for the weeks preceding the tupping season. But even then some restless spirits escape. Often Davies comes up to me and says, 'That flamer with the little horns what you and me fetch from Dolgelley last year have gone. He do go over Tryfan side when he have escape.'

And off John Davies goes with Bett and Jim to seek the erring male. There are violent fights too. I paid four pounds for a ram one year, only to find him dead the very next morning. Once we missed a ram, and found him hanging by his horns from the top wire of a strand fence which he had failed to jump. But he lived, The very end of October comes, and we gather down the ewes. The flock has shrunk. The four-year-olds are sold; the wether lambs are sold; the ewe lambs are at wintering. I always conduct a winter dipping. When I was first at Dyffryn the sheep were swarming with ticks. In spring the white, short-wooled lambs were crawling with them as the parasites transferred their attentions from mother to offspring. This harassed the lambs so that they visibly lost condition. But one year I tried a winter dip on the ewes. This is not arsenical like the scab dip used in summer, but it contains carbolic. Yearly thereafter the incidence of ticks shrank, until today the flock is completely clean, and now we see only two or three ticks in twelve months.

And this dip waterproofs the fleece, so that very few ewes lose their wool. The trouble and money is very well spent. As we dip we dose the ewes for liver-fluke. The fluke is a parasite which lives on a sheep's liver, until finally the animal dies. The fluke has a curious life-cycle. The adult lays its eggs on a certain type of small water-snail, which is thus forced to become the host to the parasites which presently feed upon it. The snail makes its home upon grasses in marshy places. The sheep eats the grass, and with it goes the snail. Thus the fluke is introduced to the sheep. For a little the sheep is stimulated by the titivation of its liver, but quite soon afterwards it will lose condition and die. In wet seasons I have heard of farmers losing one sheep in three through the activities of fluke.

There are two ways to fight fluke. The first is to spray all damp places with a weak solution of copper sulphate. This destroys the snail host who is necessary to part of the fluke's life-cycle. But on a place the area of Dyffryn spraying is utterly impracticable. So we use the other method, and dose every ewe with one cubic centimetre of carbon tetrachloride. At first we used to give this dose in gelatine capsules, but then we discovered an ingenious instrument which dosed the sheep with a combined fluke and worm mixture. A half-gallon container is strapped to the operator's back, which is connected by a flexible tube to a pistol-like apparatus in his hand. The pistol has a trigger, which, when pulled, pumps the correct dose through a nozzle. It is simplicity itself to slip the nozzle into a sheep's mouth and to shoot in the liquid. This careful annual treatment must have fatally interrupted the life-cycle of many a happy fluke. And now the incidence of the disease has shrunk to nothing.

Each year the ewes are dosed and dipped, and then the whole medicated flock is turned to the ffridd. A day or two later, perhaps on the 4th or the 5th of November, we select a dozen rams to join them. These are the pick of our males, and are therefore allowed their pick of the ewes. At the same time our homebred rams, six or eight of them, are sent up to the mountain to seek out any ewes who have not come in at the gathering. A week or so later the rest of the rams are released.

A ewe comes into season at intervals of three weeks at this time of year, and some farms have a ffridd which is inadequate to keep the ewes for the six weeks necessary to ensure that every one has taken the ram. So the flock has to be left upon the mountain, and the rams are turned up to it. But the flock is much more widely dispersed over the greater area, so that a higher percentage of rams is needed. And rams have a habit of collecting a harem of half a dozen ewes whom they serve repeatedly, to the neglect of other ewes. In the ffridd we can combat this leaning towards monogamy by gathering the sheep daily into three or four different lots. In the confusion the little cliques are well shuffled, and every female has a fresh chance to win her man. This procedure is difficult on the mountain, and the ram percentage must therefore be increased.

We usually turn about thirty rams into Dyffryn ffridd, and they deal with twelve hundred ewes or more. Some people tell me that this number of rams is too many. I know more than one farmer who picks a hundred good ewes to be shut in with his one best ram. The results are always excellent, but in these cases the breast of the ram is smeared daily with red raddle. Thus he marks each ewe which he jumps, and she is removed from the enclosure. This method would involve such labour as to be impossible in a large flock. And my proportion of males to females cannot be far wrong, because we find each spring only a handful of barren ewes.

And gradually the rams become less restless. By December few of them trouble to climb to the top of the ffridd. The younger ones, undeveloped in stamina, have ceased to work altogether.

Davies comes to me a week or two before Christmas. 'It be time to get them rams in and let the ewes go up,' he says. And we gather the ffridd to separate the weary, foot-sore warriors, who look as though they will in truth need the next ten months to recuperate. The ewes stream back to their various patrols upon the mountain. Dyffryn enters upon its most quiet period, for we do not gather again until the end of March, when the lambing time will be almost on us again.

23

The Hydro-Electric Set

FOR ENDLESS DAYS THROUGHOUT MANY winters Esmé and I had battled against the rain which our hills would maliciously call upon to discomfit us. We were used to seeing the valley change from lazy placidity to a maelstrom of conflicting elements. Dry gullies would become streams; in a few minutes, streams rivers, and rivers lakes. For Dyffryn has seven times the rainfall of London, and four times that of Bettws-y-Coed, six miles away. The Japanese have made a national cult of judo, which means the 'gentle way'. Opposing strength is turned to confound itself, so that the uninstructed opponent is defeated by his own ferocity. The weather had taken toll of us many times at Dyffryn. Could we not use its very exaggeration to our advantage? The Idea came to me. I spoke to Esmé.

'Would you like an electric cooker, an electric heater for the water, electric fires, electric light, a vacuum cleaner, a refrigerator?'

Esmé went into that daydream in which women indulge when house improvements are talked of. She spoke of the cleanliness of electricity, of its convenience, of the difference it makes to one's standard of living. Then she woke up rather cross and told me not to make her discontented. But I broached my Idea.

'I was a fool not to have thought of it before,' I said, not meaning it. 'We have one of the heaviest rainfalls in the country, and are in an ideal position to harness the water. We need a reservoir, a pipeline downhill to give pressure, and a shed for the turbine and dynamo. The power won't cost a penny once the set is in. We'll save coal and paraffin. The cottages can have free light. And it will be a blessing in the buildings in winter.'

Of course, we bundled out of the house there and then. It is odd that this first survey, although later we discussed many alternatives, laid down the exact lines which the installation eventually followed. Fifty feet above the house was a hollow in the steep hillside. Really it was not so much a hollow as a slightly saucer-shaped little plateau of about two acres. At the far end the mountain, of course, sloped down to form a barrier, but opposite the mountain at the side above the house was a natural barrier of mother rock about ten feet high and seventy yards long.

'This place is just made for a reservoir!' I said. 'We've only got to run two peat banks from the rock-wall along the edges of the hollow until they merge into the hillside, and the thing is done. And we will dig the peat from inside the hollow, so that the capacity will be increased.'

We stood on the rock barrier. The drive from the road to Dyffryn house wound underneath us. At its junction with the highway two hundred feet below us stood a stone and slated cartshed.

'We'll have a straight run for the pipeline down to that shed,' I said. 'We'll fix up the set in there, and the power-house will be central for the house, the cottages and the three lots of buildings.'

A hundred yards or so to either side of the hollow two streams slid and leaped down the mountain.

'We'll cut a leat to each stream and divert them into the reservoir,' I planned. On that particular day the streams were foaming. There seemed enough power to light a large town.

'Oh, I don't think we'll need more than one!' said Esmé. We remembered that remark months later.

I wrote at once to a firm of engineers in London. In a few days their representative called. He was a typical, professional towns-man. He was at once upset by the rough atmosphere of Dyffryn. He disliked scrambling about among the rocks. He thought of the two dams which we should have to make at the sides of our hollow in terms of concrete. I told him that a peat bank would do equally well at perhaps a twentieth of the cost, and he politely disagreed. But in Wales peat dams have held water from generation to generation. We prospected the route for the pipeline. Our engineer noted the outcrops of rock through which it must pass. He talked of motor-driven air compressors, drills and blasting. I told him, in all sincerity, that John Davies, with the Celt's intuitive knowledge of stone, could carve his way through in no time with a hammer, a chisel, and a crowbar. But the engineer thought I was joking at him and became curt. At last I showed him the two streams which I proposed to divert into the reservoir. The weather had been fine for a few days, and the watercourses had dwindled to a trickle.

'You couldn't turn a turbine with these,' said the engineer. 'That's why I want a reservoir,' I explained. 'The storage will tide us over the dry spells.'

'The streams increase after rain, then?' 'Oh, yes! Enormously.'

'How much? Twenty-five per cent?'

'Much more. At least a thousand per cent, I should say. More than that. I shouldn't be surprised if they increased fiftyfold.'

And that information finally finished his patience. He did not believe me. He thought me mad, a liar, or a practical joker. And, above all, he did not understand Dyffryn. He had tea with us, and his preliminary opinion pricked the bubble of Esmé's wildly excited optimism. He thought he could recommend a small lighting set, no more. His written report was much delayed. After some weeks I wrote to his firm. They were sorry, but their representative had been taken ill on his return. And the final report, when it came at long last, was most lukewarm.

I looked elsewhere. I found a small Welsh engineering concern, a father-to-son family affair, which lit the village in which it was located by its private hydro-electric station. The workshops were a collection of wooden sheds, and the sons, who ran the business, were slow-moving, like farmers. They had none of the crispness of professional men. One of them came to Dyffryn, I let him make the suggestions. Yes! The hollow was the place for a reservoir; two peat dams from the rocky wall back into the hillside would hold water. The cartshed was ideal. It would save building a power-house; it was central, it was beside a stream into which the water could be turned when it had given its impulse to the turbine. There was rock in the line which the pipe must take, but my new man was a Welshman. He made light of the difficulties. He pointed out how the grain of the various outcrops ran, and explained how easily the stone could be chipped away. I asked for an estimate and for a computation of the amount of electricity which could be generated. The streams were again low that day, but he knew the climatic conditions of the district, and could visualise the amount of water which would come down when the rain came.

In a few days I received his quotation. He said that five kilowatts could be generated at 250 volts. And his estimate for doing the skilled work was very reasonable.

The interest on the capital outlay would provide us with light and power at an infinitesimal part of a penny per unit. But my new adviser did not wish to contract for making the reservoir, for digging the leats, nor for digging the trench for the pipeline. This did not worry us, for we agreed with him that we could probably use our own labour and finish the work at less cost than he dared quote. However, I was not satisfied with the amount of electricity. I had set my heart on at least seven kilowatts. This amount would give us scope for a really large cooker. We had further correspondence. My electric people were unwilling to install a plant for more than five, but would do so on my responsibility. However, they made a very constructive suggestion. There is a governor fitted to these plants. At peak load the full jet of water is, of course, allowed to spray on to the spinning turbine; when the load decreases because, for example, a heater has been switched off, a contrivance deflects a proportion of the water below the turbine, so that it slows to keep the voltage constant at the now reduced load. But with this system the water is running always at full pressure, though the governor varies the amount which impinges on the turbine. The new suggestion was that I should have fitted a type of governor which would regulate the flow of water through the pipe. This would have the same effect on the speed of the turbine, but would save precious water in the storage reservoir. For when the set was running on a light load, the flow of water would be greatly decreased by a pearlike contrivance which would slide into the nozzle of the pipe.

This suggestion heartened us, for we knew that we were being ambitious. But I was still afraid of putting in a large plant for which we would have only enough water for an hour or two's running each day. So I prospected along the hillside towards the little Glyder. Within three-quarters of a mile were five more streams which could in the utmost necessity be collected in one long diagonal leat. The leat would have to run through many difficult places, but its construction would as a last resort be possible. We ordered the seven-kilowatt set, and our engineer came over to help us to lay out the course for the pipeline and to determine the levels for the dams.

He brought with him only two instruments – a spirit-level and a twenty-five-yard tape measure. We went up to the hollow, and planned the foundations of the two dams so that they splayed outward from the rock-wall as they ran back towards the slope of the hill. The inner tips of the dams would thus be about a hundred and twenty yards apart. It was important to build them up to the same height and, as the bed of the hollow tilted a little to one side, the heights of the dams above ground-level must differ if they were to match at water-level. My engineer surveyed the levels with a spirit-level laid on a ruler, along which he sighted his eye. I drove in long bamboo poles, once bought in the hope that the wind would countenance sweet peas, as he shouted directions to me across the little plateau. Owing to the tilt of the surface we found that to match a depth of eight feet at the lower side, the dam on the opposite side would need to be only four feet above ground, for the ground at that side was four feet higher than at the other. I determined to dig all the materials for both dams from the higher side, so that the floor of the little lake would level up. This would have the effect of sinking the bottom of the lake at the higher side below the ground by some four feet. Then we plotted the course of the pipeline down the very steep slope to the cartshed. Several outcrops of rock showed in our path, but with a fine carelessness we ignored them.

On a cold November day we cut the first sod of Llyn Dyffryn. We began by slicing off the turf over the higher half of the plateau. We built up the turves into the four-foot wall necessary to retain the earth of the lower of the two dams. I had taken on an extra man for the work in front of us, one Len by name, who had worked at casual labour for me over a period of years. He was English, lived at Llandudno, and was a gregarious soul with a liking for snooker halls. He was a queer body to become a suitor of Dyffryn in her austerity. But he could never keep away for very long, and when work could be found for him Len would cheerfully forsake the smoky lights of the billiard saloons.

The work went well at first. Quickly we removed the turf over an area of fifty yards by thirty-five, and piled it into a neat wall. The layer below the turf was peat. It was as easy to cut as butter, and we took it out in tidy bricks which we built four or five feet deep against the turf wall. As we built we stamped the peat well down, so that each brick merged into its neighbours, and no crevices were left through which the water might find its way. We dug farther from the dam day by day, and soon had to call upon Belle to help us with the cart. But our quick progress was short-lived. At a depth of a foot or so the gentle peat changed to hard-packed clay. The clay was greyish in colour, and was mixed with gravel. It was so hard that we had to chip it away with picks, and it broke reluctantly away in the smallest lumps. This was a great setback. We had to dig deep, not only because much material was wanted for the dam, but because the pipeline had to draw from this end. And the entry of the pipe should be at the lowest point of the lake in order to give the lake its maximum storage value.

So we chipped away doggedly. The clay was grand stuff for the dam, for it set in as hard as concrete. But it nearly broke our hearts as we scratched at it week in and week out in all weathers. And the winter weather became vicious. We had a little snow, some sleet, some hail, much rain, and sometimes hard frost which congealed the ground to iron, and gave us a most welcome excuse for a rest. The pit which we were gradually excavating was always half full of water. The clay became glutinous when loosened. When one tried to fling a shovelful on to the dam, the clay stuck to the tool, and it became a hard-swearing, back-breaking business.

Meanwhile my engineer had sent a factotum to wire the house, the cottages and the buildings, and to commence work on the power-house. Daily material began to arrive. First came a dynamo. It was not enormously big. The lorry-driver asked for help to unload.

'Come on, dad!' cried Thomas. 'I'll lift it down to you.'

He gave the dynamo a push in the direction of the lorry's tail-board, but it did not slide. Then a harder push, and again a harder, but there was no movement.

'You do want to give it a good shove, man!' grinned John Davies. 'Not potch about.'

'She be bolted to the floor!' panted Thomas, indignant at this slight on his manhood. But the dynamo was not bolted down. It weighed seven hundredweights. The lorry-driver gave a cynical smile, and rigged up a little derrick to lower his load quite easily and gently to the ground. After the dynamo came sand and cement to concrete the floor of the power-house, then the governor to regulate the water-pressure, then the Pelton wheel or turbine. And later there arrived lorry-loads of six-inch pipes. The top half of the pipeline was to be of reinforced concrete pipes, and the lower half, where the pressure would be increased, of asbestos cement pipes. Coils of thick copper wire were sent for the overhead power lines, and creosoted poles to carry the wires. Down below everything went well. The factotum proved a universal genius. He wired the house, cottages and buildings so neatly that there was little or no mess to clear up. He made an elaborate concrete design on the floor of the power-house, sank the Pelton wheel into its pit, led in the six-inch pipe which would eventually carry the water, and connected the governor with the control in the one-and-an-eighth-inch nozzle. Then he set the dynamo and connected it with the turbine.

All this took some weeks to complete. Up at the lake we had worked just as hard, but had less to show. The one dam was finished, and looked imposing from the inside as it towered over our heads. We had started on the other dam, a little appalled at the height of the eight feet above the ground to which we must raise it. We pulled down some old stone walls near by, and John Davies built a five-foot retaining wall from the rock barrier towards the slope. Naturally, as the wall reached farther and farther towards the mountain, the ground rose, and the wall, level-topped, shrank in height, till after fifty or sixty yards it dwindled to nothing. We intended to cut a channel at the end of that dam to carry off overflow. And then we began to bank our peat and clay against the wall. We dared not make the base of our dam less than ten feet thick, and a cartload was swallowed up on the mound as if it were a spadeful. This laborious work went on slowly, but at least surely, until the ground about the new dam became badly cut up by the horse and cart. Belle sank several times a day, and at last it became impossible to approach closely. We had to tip our loads at increasing distances, and to throw them by shovel. We tried making a road over the jellied peat, but we had already used in the wall most of the nearby stone. Then we cut a way down through the peat to the firm clay, but this trench filled with water. In the end we made a track of planks and wheeled our filling little by little in barrows.

We worked on that lake through the worst of the winter, and February came before we finished the second dam. But when it was done it looked very fine. At once we began on the trench for the pipeline. We had left a gap through our first dam for the pipe's entry to the lake. The floor of the lake at that side being four feet below ground-level, we had to sink the pipe deeply for its last stage. We entered the pipe, clamped on the grating through which it would draw, then we closed the gap. There was not place for more than two to work efficiently on the trench, so the other pair began work on the two leats which were to cut across the hillside from opposite sides of the lake. Much rock was encountered on both the trench and the leats, but at last the pipe could be laid, and the leats, one a couple of hundred yards long and the other a hundred, were separated from their respective streams by only a thin barrier of uncut ground. The concrete pipes were in lengths of five feet, and were heavy. It was unfortunate that they were to be used at the top. It took the combined efforts of the four of us to drag them up towards the lake, which was three hundred yards off and two hundred feet above the road. But the asbestos pressure cement pipes for the lower half of the line were lighter and easier to handle. The factotum borrowed Len to erect the poles for the transmission lines and to help him stretch his wires, and the engineer himself came down to superintend the jointing together of the pipes. In a few days all was ready.

One March morning we hacked away the last five feet of earth at the end of each leat, and we dammed the two streams. The water gurgled off to our lake, swirling with it loose earth, so that the advance-guard looked like coffee. It had been raining for a spell. The streams were fairly swollen, but even so by nightfall we had only a couple of feet of water in the reservoir. This pleased us. It showed that the capacity when full would be very considerable. At dusk the factotum opened the valve in the powerhouse. The water sprayed out of its narrow jet on to the Pelton wheel with unbeliev-able force, for there was a fall of two hundred feet behind it, and the capacity of the six-inch pipe was constrained at the end to hiss through a nozzle of just over one inch. The turbine revolved. The dynamo turned with a mounting hum. The governor came to life and took charge of the voltage. Esmé and I were waiting in Dyffryn house, where the switch of each dead light was turned down. The cottages too were keeping bated vigil. It came – a faint glow of the bulb filaments. Brighter! Brighter! A steady, flicker-less brilliant light flooded the house. The powerful outside lights lit like day the terrace in front and the yard behind. The old farmhouse too blazed from every window, astonished, a little hurt, at its enforced moder-nity. We ran to the top of the hill. Cottages and buildings were recklessly flinging bright shafts into the night. Away in Capel Curig they looked up to the Glyders and saw Dyffryn a flaming portent in the sky. It was all very dramatic.

Next morning Esmé and I scrambled up to the lake. More rain had fallen in the night. The creaming leats had filled the lake, so that waste poured round the overflow. But tragedy faced us. From a thousand points jets of water spurted through the two dams. Once the streams died to normal, our reservoir would empty like a sieve,

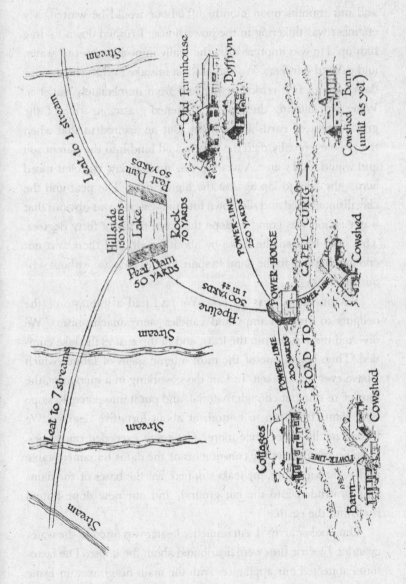

and our months upon months of labour would be wasted. My engineer was tinkering in the powerhouse. I rushed down to drag him up. He was impressed by the really imposing sheet of water, and calmed our fears. We had made a mistake in throwing up our dam without first removing the turf from immediately under it. Water was filtering through the flattened grasses in spite of the great weight of earth above them. But he assured us that when the grass eventually died the dam would bind into the parent soil and would hold water. A wind sprang up that day. Wavelets raced across the lake to lap against the higher dam. The peat and the clay disintegrated and slid down into the water. It was obvious that water action was going to slope the dams to about forty degrees. This would widen the bases by about a half, and there was not enough material in the dams to spare for this increase without serious weakening.

But our blood was up now. We had had a foretaste of the delights to come. Lamps and candles were anachronisms. We diverted the water from the leats, and in three days the lake emptied. Then began one of the most intense spells of labour which I have ever undertaken. In four days, working in a quagmire, the four of us dug out enough material, and put it into place, to slope both dams from top to bottom at about forty-five degrees. We turned in the water once more. The result exceeded our hopes. Even on rough days the constituents of the dams remained stable and, what was more, the leaks stopped. For the bases of the dams now protruded into the cut ground, and our new slope bound itself into the earth.

Our cooker arrived. An immense heater was fitted to the water cylinder. Electric fires were distributed about the house. The factotum connected our appliances with the main fuse box, and Esmé turned on the grill of the cooker. This was the first time that a serious load had been thrown on the set. The grill glowed brightly, but something was wrong down below in the powerhouse. The voltage began to 'hunt'. The lights grew dim and bright, alternating as if a pendulum were controlling their intensity. Again we sent for our engineer. He brought with him a flywheel which weighed two hundredweights, and we found room to key it on to the shaft beside the turbine. The result was satisfactory. There was no more than a quick flick when fires or the cooker were switched on.

We entered into a Golden Age. We had, for the practical purposes of our establishment, unlimited power. We had an omnipotent servant whose wages decreased the more was demanded of him. And he revelled in hard work. But one day I was walking up the hill. The pipeline, as yet naked in the unfilled trench, ran up the hillside near by. To my horror, a great geyser of water suddenly shot high in the air. A pipe had burst. The waterspout was spectacular. Several motorists stopped on the road below, and no doubt thought it a remarkable natural manifestation. As for me, I grabbed an armful of sacks from the house and ran up to the lake. Davies had materialised out of the ffridd and was busy diverting the leats. I plastered the sacks over the grating of the pipeline, where they were tight held by suction. Esmé, unutterably distressed, her lunch half cooked in the glittering new oven, phoned the engineers and blurted the story of the tragedy in broken words.

The engineer arrived next day. He had brought with him a cunning device. First he explained how the burst had occurred. Under heavy load, such as when several parts of the cooker were in use, the water was rushing at great speed down the pipe. But suddenly part of the cooker is switched off. The governor instantly regulates the flow to the more modest demand. A check runs up the column of water, and the upward and downward forces come to grips just halfway down the pipeline. While we inserted a new spare pipe in place of the burst one, our engineer fixed a valve to the vital halfway point. Should this water-hammering occur again, the surplus force would expend itself in a violent jet through the spring-loaded escape valve.

Again all was serene, and Esmé all smiles. But the boisterous spring passed, and summer came. The rainfall lessened. Our leats dwindled to trickles. Anxiously we watched over the shrinking reservoir. We began to turn off the set at bedtime to conserve water-power for cooking. We had electrified everything, and were completely dependent upon our new servant. And at last the two streams which we had tapped became quite unable to cope with the demands of the pipeline. There was no help for it but to lengthen the westerly leat along the side of the Glyders in order to enlist the aid of more streams. We went three hundred yards and captured a third stream. But again in dry spells our water failed us. We carried on to a fourth, to a fifth, to a sixth. Sufficiency was not far away. The leat twisted like a snake on a slightly upward slant across the ffridd. We dammed each successive stream so that at low water all its contents poured into the leat, but so that in flood the bulk of its water would spill over the dam and so continue down its wonted course. For had we collected all the water off that hillside during a rainstorm, not only the lake but Dyffryn house would have been washed away. Ingoing and outgoing nearly balanced now, but for good measure and to give Esmé peace of mind we cut onward out of the ffridd to collect a last stream on the mountain itself, three-quarters of a mile from the lake.

It is odd to stand up there on a cold, wet, stormy day and to think that the icy water racing away from one's feet is providing one with a hot meal, a steaming bath and the warmth of radiators, in a house nearly a mile away. Yes, we have certainly scored a point over the weather there. But no doubt she has a new weapon to use in the next round.

24

The Renting of Cwmffynnon

FOR YEARS, I, AND LATER ESMÉ, had cast covetous eyes on the great hollow of Cwmffynnon. Cwmffynnon joins Dyffryn all along the western boundary. It is nearly a thousand acres in extent. It meets Dyffryn at Gldyer Fach, and its upper boundary runs along to Glyder Fawr, whence it sweeps down a horseshoe-shaped ridge to the Llanberis Pass. The place is wild and beautiful, for the huge basin is formed of great steps, each one a formidable cliff; heather and whin are plentiful – too plentiful for sheep – lower down, and on the heights grows sweet, short mountain grass. But even the heather and the whin become sheep food when snow comes, when nothing shows but the sprigs of the bushes. Nestling like a jewel, glittering with the hard light of a diamond in the enfolding hills, lies Cwmffynnon lake. Many tumbling streams feed it on its north shore, and at the other side it gives birth to the Avon Gwryd. Cwmffynnon is cruel, beautiful, fantastic, and at night eerie.

The place marches with Dyffryn. Many Dyffryn sheep stray there in easterly weather, and whenever we gather Dyffryn we make a wide sweep into Cwmffynnon territory to collect wanderers. For many years it was rented from the large estate which owns it by one of a family from near Llanberis. Other relatives of the tenant held adjoining land, and the whole clan helped one another at their respective gatherings. The people have always been very good neighbours of mine, and we at Dyffryn have done our best to work in with them. One day not long ago John Davies came up to me.

'Pricey Pandy have tell me he be giving up Cwmffynnon,' said he, and immediately fell to filling his pipe, as if nothing else in the world interested him. Price Williams, Pandy Farm, was then the tenant of Cwmffynnon.

'Oh, is he?' I asked casually, and maliciously turned the talk to something else. Davies became acutely uncomfortable. I knew quite well what he was about. He must have met Pricey recently. There would have been conversation in Welsh.

'I am giving up Cwmffynnon,' Pricey would have announced. 'I be sorry to hear that,' Davies would answer. 'Us have been good neighbours.'

'You might tell the boss,' from Pricey. No more would be said, and no more needed. Davies would know at once that Pricey wished me to take the tenancy from the estate, and to agree a stock valuation with him direct. For there were reasons why I should be acceptable to the rest of the clan as an even closer neighbour. For one thing we had always worked in harmony, and again Dyffryn was capable of giving powerful interchange of help at gatherings – for as tenants of Cwmffynnon we should need the clan's help to gather. And Cwmffynnon was a remote place. Many sheepwalks circled its boundaries, most of which belonged to Pricey's family, so that large numbers of strange sheep would always be expected there. It would be a relief to the clan to know that no dishonest work would go on, hidden by the hills which screened the hollow. But Pricey would never approach me direct, nor would Davies as his emissary. There is nothing despicable in this circumlocution. In long-subdued races it becomes a mannerism, and a mannerism is bearable so long as its cause is understood. The Welsh are always circuitous in their dealings, but they are at least as honest as other nationalities.

Almost immediately I left John Davies, who was uncertain whether his message had been understood by me, and drove down to see Pricey Pandy. Pandy Farm has a little symmetrical house deep in the pass of Llanberis. The sun, hidden behind the towering sheer north ridge of Snowdon, only touches the house for five months of the year. For the rest of the time the shadowed dwelling broods among the boulders, and its blind windows stare at the toppling hills. The approach is up a stony track creased by coursing water. The track is too narrow to take a horse and cart, and generations of tenants have humped home their meagre provisions on their backs. The track peters out in a field, hard won from the mountain whose rock ribs stick through the turf. As I reached the field a spate of dogs poured from the open doorway of the little farmhouse which stood at the opposite side. Luck and Mot raced ahead, and a conflict of patrols followed. I ran to stop the battle, and was aided by reinforcements from the house – Pricey himself, his brother Jonathan and his uncle Idris. Presently they led me inside. There was tea laid out on the kitchen table, and an ancient woman hovered in the shadows. Tea is in progress in Welsh farms from lunch-time until supper. As we ate we discussed prices, the damnable agricultural policy of the Government, little Davydd Pentref's drunken encounter with a bus-conductor, a rumour that two Dyffryn yearlings had been heard of at Ebenezer, behind Llyn Peris, and sundry other important topics.

'I'll go and hunt for the yearlings as soon as I leave you,' I said. 'Pricey will go with you for the ride,' said Uncle Idris.

'These strays waste a lot of time. There are a good many of our Dyffryn sheep in Cwmffynnon too.'

The taboo had been broken! Cwmffynnon! The three men gazed uneasily at one another. The momentous atmosphere became almost tangible. The wife of Uncle Idris in her agitation swung to and fro the crane which suspended the kettle above the glowing peat. The two china Pekingese on the mantel held their breaths. Brother Jonathan dropped his pipe, and only a miracle saved it from destruction on the stone flags. Pricey stirred his tea till the strong liquid swirled over into the saucer. And at last Uncle Idris said in a voice whose emotion mocked his elaborate unconcern, 'Pricey be giving up Cwmffynnon.'

My memory slipped back along the corridor of the years. It seemed long ago since the storm had blasted me through the back door of Dyffryn. Sitting in the little low dark kitchen at Pandy, surrounded by the spare dark men of the place, I could recapture without effort the picture of that earlier day when I had blindly fenced with the old farmer and his wife. I had known risk at Dyffryn, death, and love too. I had lost money, made money, been beaten to the earth, and had learned to get up again. The boy had, I hoped, become a man. And I asked no more.

'There's good grazing there,' I said. 'Well grand!'

'How many ewes are up there, Pricey?' I asked.

'About a hundred, and thirty-five yearlings be wintering near Caernarvon.'

I knew that Pricey had been selling heavily of late, but had had no idea that he had depreciated the numbers of his permanent flock to that extent. For Cwmffynnon could carry quite four hundred breeding ewes, and in summer, when the lambs had been born, and when the yearlings were home, there should be eight or nine hundred head of sheep there.

'Is that all?' I exclaimed. 'It'll be years before I can build up a proper-sized flock.'

Cwmffynnon is, of course, an open mountain, and it is therefore impossible to purchase strange sheep to help stock it. I knew, though, that all Pricey's sheep were young. And they had had less attention than they should for some years, so that only the fittest had survived. There were no sheep more healthy and vigorous in Wales – and none so wild. So I hoped that it would be unnecessary to sell off any draft ewes for three years or so, their vitality was such that they would thrive on the mountain up to five or six years of age. I worked out in my head the increase I could hope to gain in three years. It was now November, and there were a hundred ewes on the sheepwalk, with another thirty-five yearlings wintering. The hundred ewes would bring eighty lambs in April. Forty of these would be ewes, and would a year hence be wintering in turn. But the ewes, augmented by the present yearlings, would the next November number a hundred and thirty-five, and would produce, say, a hundred and five lambs. Fifty-two of these would be ewes again, so that by the second November there would be a hundred and thirty-five plus the forty yearlings, or a hundred and seventy-five ewes for the ram, and fifty-two yearlings wintering. After this I should have to reckon on selling off some of my ewes each year, for they would be getting aged, but the annual entry of yearlings into the flock would more than compensate for the sales, so that my increase would slowly continue. I thought that the 400 breeding ewes which I wanted would be reached in six or seven years. And then, of course, the profits would start, for I could each year sell off an equal number of ewes to balance the entry of the yearlings. Meanwhile I would have half my lamb crop, the wether lambs, to sell annually, and there would be the wool receipts. Expenses would be small. The rent of Cwmffynnon, isolated as it was and without a house or buildings, was extremely low, for it was virtually unworkable for anyone except myself or the Pandy people. And I should need no extra labour. The only items would be wintering costs, rent, and such small items as dip. Although the present flock was disappointingly small, I determined to take the long view and to negotiate, for in conjunction with Dyffryn I knew that Cwmffynnon must in the end be a little gold-mine.

'What about ingoing?' I asked.

'The estate have had a twelve-month notice from Pricey, up next November. But the agent have say if Pricey do find a tenant he can go now.'

'Do you want a valuer for the stock?'

'Well, I do believe that Pricey and you can come to something between the two of you.'

Pricey nodded agreement.

'All right!' I said. 'When can you gather?'

'Next Saturday,' spoke up brother Jonathan. 'In our places by ten o'clock.'

Some of the clan were only part-time farmers who also worked in the great Llanberis slate quarry – one of the largest in the world. It suited them to gather on a Saturday. I went back home and told Esmé the news. She bubbled with excitement. We got out a map, and before long had rented or bought the whole eight-mile range of which the two Glyders formed the eastern end. And by bedtime we had arranged the working of the huge block of sheep-land. It was difficult to come down to earth and think of just Cwmffynnon.

Saturday came. John Davies, Thomas and I left at eight and climbed in a long slant up to Glyder Fach. We sat a little and had a smoke, while the dogs sniffed the ground and tensed their ears. Cwmffynnon opened at our feet, and deep down its lake glistened like a drop of water in the bottom of a bucket. Glyder Fach, where we sat, and Glyder Fawr, along the ridge, were the highest points on the crescent wall which formed the hollow, and which graded down to touch the Llanberis Pass at its highest point of about 1,200 feet. Now and again we caught glimpses of tiny figures moving up this ascending ridge from the pass. The Pandy people were climbing to their places. John Davies and I presently moved off along the ridge towards Glyder Fawr. Thomas stayed where he was. In a little

I too stopped, and Davies went on alone. Ten minutes later Bett made contact with Pricey's Prince. We all sent our dogs barking down the precipitous hillside and slowly followed them. And as we gradually neared the lake, the line constricted until we could shout greetings to one another.

There is no place so difficult to gather as here. The sheep had been left long untended, and had reverted to the wild state. The numbers were far too few for the big area in which they lived, and the scuttling groups of two or three ewes did not meet and merge. So we had to treat individuals and couples quite separately, because there was no flocking together. Among the boulder-traps, innocently covered with heather or whin bushes, men and dogs floundered hopelessly, while the sheep, who knew every track of the ground, and who were too wild to be intimidated, broke through the line as they wished. Sometimes, when hard pressed, little troops would take refuge on giddy ledges of the many cliffs, where the danger to sheep and to dogs forced us to leave them. Goats abounded in these striated cliffs, and their pungent smell distracted still more the frantic dogs. Luck and Mot were exhausted by the time we reached the lake shore, and with lolling tongues they paddled belly-deep into the icy water to revive their energies.

It looked to me as though we had taken a sorry catch. We had used a sledge-hammer to drive a pin. The handful of surprised sheep who were snorting and stamping within our circle was a small reward for the prodigal energy spent by eight men and twenty dogs. By the side of the pass are some ancient stone pens. We drove our flock into them, and I found that there were more sheep than I had thought. But owing to the small numbers of the Cwmffynnon flock I knew that many strangers would be encroaching on the grazing. And, sure enough, we picked out sixty or seventy ewes from various sheepwalks of the Pandy clan. Then we counted the Cwmffynnon ewes. There were only sixty-five there; all three of the rams had come in. But I did not doubt that Pricey had a hundred ewes on the place, for I had myself seen sheep breaking back in all directions. I knew well the Cwmffynnon sheep, and had no need for a close inspection. The men withdrew, and Pricey and I were left alone. There was no attempt by either of us to put across a smart deal. I think that we both asked no more than to agree a fair price.

'Well, how much?' I asked.

'Forty for the ewes and twenty-five for the yearlings,' said Pricey.

'I won't argue much about the ewes,' I answered, 'because they're all young sheep. But you want too much for the yearlings. I'll give you thirty-eight for the ewes and a pound for the yearlings. And I'll buy the rams at fifty shillings.'

And quite soon we had settled at thirty-nine for the ewes and twenty-one shillings for the yearlings. We were not very far out in our valuation, for the draft ewes at Dyffryn sale had fetched an average of thirty-six shillings. The difference in price was a modest acclimatization value. It was rather a dear market for an ingoing – I had gone into Dyffryn at twenty-six shillings – but I wanted the place, and there were then hopes, since lost, of a boom. Pricey and I struck hands; the others crowded round agog. Uncle Idris drew me aside.

'We will do our bestest for you!' he whispered. 'Ask us for help when you want her.'

I thanked him, and went to help stamp the bought ewes with pitch, for, according to custom, I had agreed to take at the same price all other sheep which should come in later, and we had to distinguish between these and them. Quite soon a car stopped on the pass above the pens, and Esmé came scrambling down to us, eager to view our new flock. She was disappointed at the number, but was enough of a daydreamer to be able to visualise these same pens some years later, crowded with sheep, bustling with activity. We turned the flock out to find their way back to their home, while the Pandy people went off down the road with the strays. Each man would turn his bunch on to its proper mountain, but I had no doubt but that the wanderers would be back on Cwmffynnon within twenty-four hours. However, the yearly increase of the native flock would gradually force back these invaders.

Then we Dyffryn dogs and people piled into the car, and Esmé drove us home. Cwmffynnon gatherings are still very difficult, for there has not been time for an appreciable increase of stock as yet. But the time will come when constant attention will take the edge off the wildness of the sheep, and when thickening numbers will make the present scattered units act as a flock. But the vitality of these sheep is beyond price. I intend to keep as many lambs as possible to use for rams in the Dyffryn flock, for the strain will have an excellent effect, and will round off my hardening process.

The renting of Cwmffynnon meant more to me than the acquisition of land and stock. I dropped that day the keystone into the arch of my capacity. I felt that though I must continue learning till I died, yet I had graduated, I knew now the extent of my knowledge, and had taken the measure of my ignorance. The fledgling could take wing. The flight might be shaky, but the goal was known. And in the hills one climbs a range to see another beyond; one learns how relative is ambition. The sense of values too develops, and this is the most important sense we have – to know what is worth while and what is useless, what is real and what is false. Since Esmé joined me we have lived six years together at Dyffryn. At first we built the structure of our life with unknown materials; some proved strong and enduring, others gave way under strain, and left us to prop frantically our leaning edifice. But in a little we grew to discriminate, and to select with fair success the good from the bad. We learned to judge people by what they are, not by what they are said to be. And we learned that no work is unworthy so long as it be well done.

But many people have no use for these abstracts. Rightly or wrongly they rate more highly the spoils of concrete gain. Of these we have less to show. We still cannot afford many things we desire. We have stinted ourselves, but we have never stinted Dyffryn, and I truly believe that one day she will repay us. By selling lightly at the annual sales the flock has expanded by some hundreds of head. This has curtailed our income, but increased our capital. We have bought the most modern mechanical equipment for mowing our hay and for cultivating our meadows. And agricultural machinery is dear. My grass-harrow cost thirty pounds, my hay-sweep thirty, my tractor mower thirty-five, my autoscythe fifty pounds. But the equipment is there, and the capital pays handsome dividends by reducing labour charges. We have improved Dyffryn house and the two cottages by structural alterations, and by fitting bathrooms in all three. I suppose this cost three hundred pounds, from first to last. And, again, the hydro-electric scheme cost three hundred and fifty pounds. We have renewed miles of fences, dug many more miles of ditches to drain bogs and swamps; we have concreted floors, put windows in dark buildings, ventilators in stuffy ones, and laid on water in dry ones; spread lime by the score of tons on acid soil, and paid for the lime; scattered basic slag; harrowed acres, or perhaps square miles, of matted pasture. Dyffryn is a costly mistress, but I do not think an ungrateful one. One day she will grant her favours. Just now she is still testing us. We are struggling now to turn our hands with facility to carpentry, brick-laying, motor-repairs, veterinary work, electric troubles, salesmanship, advertising, housekeeping. But we have fertilised Dyffryn, and Cwmffynnon is her offspring.

So the gains, such as they are, have been concrete as well as abstract.

Esmé and I are converted to the land, and we know that in this country above all others there is urgent need of more disciples.

Great Britain conceived its big population in the womb of industry. But British industry is growing feeble, and can no longer support her children. Britain was first in the industrial field, and during the nineteenth century won a clear lead in the race for world markets. She became a vast workshop for the rest of the globe. Her foundries and her factories expanded far beyond the needs of her own home market, and even far beyond the needs of her Empire. Then one by one other Western nations learned the secret of our success and crept close behind us in the race. The Golden Age of Britain faded. The competitors came abreast. The proud leader had to struggle wildly to keep up. Then the East woke from her sleep with startling speed. The pampered white man was forced to compete with the fatalistic Japanese, who worked willingly enough for a few pence a day and a bowl of rice. The white man could not stand the pace. He had either to lower his own standard of living, lose his markets or sell his products below cost. One or other of these alternatives has been accepted by most industries. Then came the totalitarian countries with their complicated economics. State-controlled labour and subsidised exports allowed them to fight the East, but the remaining white countries, the democracies, with Britain panting at their head, were elbowed out of the race. We have lost the Eastern markets which our workshops are manned and equipped to supply. These are gone for ever, and any honest industrialist will admit it. We have also lost much of our industrial trade with our own Empire. For our Dominions are yearly installing more and more plant to satisfy their manufactured needs.

The Government publish annually a statistical abstract. I do not want to quote figures in this book, but I wish indeed that our blind legislators would have someone read to them their own publication. For many years our imports have been around the thousand-million-pound mark. For the same period our exports have been less than half this figure. But the industrialist has a wonderful genie at his command. He has only to rub the magic shilling which he always carries because it was the first coin he ever earned, and the genie will materialise. He is a tiresome fellow, very abstract and windy. He calls himself Invisible Exports.

'Explain away the deficit between income and expenditure' commands the industrialist smugly.

Invisible Exports ebbs and flows about the place, wheezing and blustering. He is not very intelligible.

'Insurance... shipping... overseas investments,' burbles the genie.

'Thank you,' says the industrialist, and with a satisfied hand waves Invisible Exports back to the copper jar.

But these relatives of the genie are less amorphous. Insurance, shipping, overseas investments are neatly tabulated in the statistical abstract. One year they covered a fifth of the deficit, in no year have they covered a half. So the inquiring layman can only believe that we as a nation have been living for many years beyond our income to the tune of several hundred million a year. Of course we are, or rather we were, a very rich country, with tremendous financial reserves. And each hundred million is only about forty shillings a head. But it is only by reason of this store of wealth that we have survived so long, carrying as we do the burden of a good living standard with its attendant expensive social services. For we are unquestionably living on capital. When an individual does this he presently goes into bankruptcy. A nation does just the same – it inflates its currency.

There is no question but that our redundant working population must be dealt with on new lines. It is useless to continue to flog the donkey who lies dead between the shafts of the half-empty cart. We must remove the carcass and hitch a smaller beast to a lighter vehicle. And for the surplus workers of our inevitably contracting industry there are two courses to take. They can either go to the Empire or till the land at home. There are many grave difficulties with either alternative. It is hard to ask men to uproot themselves from home, friends, relatives and accustomed ways in order to undertake a life of undoubted hardship overseas. Again, the Colonies will not have men without capital. It is doubtful if this country could find the money to capitalise the dole and to give the sum to the emigrants. And the Colonies have had unfortunate experiences with Englishmen. Many of the wrong type have spoiled the market for their betters. I have myself seen signs: 'Help wanted. No English need apply.'

So that there is much prejudice to overcome. Of course, if we do not farm our Empire someone else will do so one day. There are already sixty thousand Italians in Australia. But necessity may do what legislation cannot. The certainty of inflation and the certainty of curtailed social services to husband the country's melting capital may drive men overseas. All emigrants are the spawn of hardship.

We eat six hundred million pounds' worth of food a year in Great Britain – at producer's prices. About 42 per cent of this food is grown at home – that is, a value of two hundred and fifty million pounds. This vast amount of food is produced by an agriculture which is as sick as it can be with survival. The farmers are sacrificed on the shabby altar of industry time after time as the industrialists throw good money after bad in efforts to thaw frozen foreign assets or to stimulate declining barter.

We buy from the United States and from the Argentine. We have bought in vast quantities from European countries whose free power to export has lain under the shadow of Germany's predatory claw. And for agricultural imports we have relied much too exclusively on the Scandinavian countries, whose foodstuffs are now diverted from us to the aggressor. And our farming is the defenceless bait which lures countries to the agreement. Even the recent trade pact with Ireland was at our expense. I well remember the shiploads of Irish cattle which poured into Holyhead when the tariff was removed. The country was flooded with them, and the price fell by pounds in a month. I lost over a hundred pounds myself. But what did it matter? Our precious manufacturers were drugged to a new spasm of life. Of course, we farmers can compete with foreign countries. But first let us understand whether or not the British standard of living is to stay or to go. For we can compete only by halving our men's wages and lowering still further our own competence. Already our living standard is lower than any other class in the country. The farm worker has only just become entitled to the dole. Officially now, I suppose, he is considered human.

But imagine this. Imagine British capital invested in British land, secure from moratoriums, wars and national bankruptcies. Imagine the farmers engaging men for their neglected work, for draining, ploughing, cultivating. There are half a million holdings in this country. There is no doubt these could absorb half a million men were farming worth while. Our present Minister of Agriculture, Sir Reginald Dorman Smith, says a million. This will not solve the problem of the three or four million unemployed who will be idle after this war, when the immensely over-expanded engineering industry has ceased its artificial war production, and when the fighting services fling men back to civil life by the hundred thousand. But it will help. Our exhausted agriculture produces its two hundred and fifty million pounds' worth of food now, while in extremis. Were it healthy it might well add another two hundred million to that figure. These need not remain dreams. They could be realities. Two hundred millions saved from imports would not balance our budget, but, like the absorption of some of the workless, it would help. And for every pound that industry invests in agriculture, nineteen shillings and eleven pence will flow back to the factories as farmers renew their worn-out machinery, modernise their antiquated equipment, expand their cultivation. This will not replace the for ever lost world markets, but again it will help.

At present the money spent abroad in food and the capital invested abroad returns but in small quantities to our starving manufacturers.

Men are loath just now to return to the land. The life is hard, the wage small, and the instinct of husbandry is dead in them. But man was born of husbandry. In the bleak times ahead he may turn again to his only sure help, the soil. He will readjust his values, and may taste in the end the ultimate joy of tending Nature in her labour.

And now dogged democracy rouses its slow strength to combat tyranny. Free men have spirit still to die for the right to live. The quiet men from the peaceful fields will march as well as any, for they have reality to defend. And when the storm fades and the dust drifts leeward, we will return to a thousand Dyffryns, proud to carry our dead.

Afterword

by Dafydd Morris-Jones

TWELVE YEARS AFTER THOMAS FIRBANK bought Dyffryn, and some 50 miles south, my grandfather bought our farm Tymawr from the Hafod estate, to which he'd been a lifelong tenant. By this time Firbank was in the Coldstream Guards, the world had been plunged into all-out war, and even in rural Ceredigion there weren't many aspects of life left untouched by the all-pervasive reach of global conflict. The farmhouse was brimming with evacuees, government inspectors advised on and sanctioned the farm's productive output, and the prices of goods into and out of the farm were pretty well fixed at national level. Rationing applied to farmers as well as those who consumed the fruits of their labour, and detailed inventories of all produce were kept, checked and cross-checked in an attempt to minimise transgressions and black-market trade. It would be easy to assume that with such transformational change and tight central control, very little of what came before would be particularly relevant to this new reality, but nothing is ever that simple.

There had been a change in the countryside air since the end of the previous War to End all Wars; small farms and gentry alike had lost descendants, lines of succession for both tenants and land-owners had been severed, generations of hard-won knowledge had been lost, and the new economic landscape of a post-war, post-Victorian world was rapidly reshaping the social and physical land-

313

scape from Llanlleiana to Trwyn y Rhws (Rhoose Point). By the 1930s, in the middle of an agricultural depression, there were an increasing number of ageing farmers looking to retire but without anyone to pass the land on to. As was the case for Thomas Firbank at Dyffryn, farms were coming up for sale that hadn't been on the market for centuries, if ever. These lands, however, were dwarfed in scale by the sale of land by the estates of the gentry over the first half of the twentieth century. Tens of thousands of acres were put up for sale over the course of a few years, most with tenants in occupation, many of whom stood little chance of being able to raise the capital to buy their homes and land. This was the situation my grandfather found himself in at the beginning of the 1940s. The government had manufactured a ready solution that met the needs of both state and vendor – the Forestry Commission would buy any land which couldn't be expediently disposed of, evict the tenants, and plant the land with military rows of uniform conifers. Without ready access to bank lending (there was a war on) and in the face of land values kept artificially high by the fallback position of state-funded purchase, this fate awaited many farms, and some of the most important grazing land in the area. My grandfather was one of the few Hafod tenants who managed to avoid losing his home to the Sitka Spruce, and with the help of an unofficial money-lender completed the purchase of our farm barely a day before the estate owner died unexpectedly and a whole new round of turmoil was visited upon the remaining properties and assets. It took him the rest of his life to repay the debt, and while doing so he watched the evergreens grow over the hay-meadows and houses of his neighbours, fragmenting pigsties, pastures, peatbogs, sheep-folds, ffridd, hefts and grazing rotations, and remained extremely thankful for his good fortune.

With a 21st-century lens, it would also be easy to assume that the changes in agriculture, wider society and the economy during the second half of the twentieth century render the influences of previous generations almost void. With the dawn of the European Common Agricultural Policy, the national focus on efficiency and maximising production through the 1970s and 80s, the consolidation of smaller holdings, the mountainous machinery now deployed on country lanes and fields (normally coinciding with peak holiday season!), the days of the Allen Scythe and the lithe fell-running farmer seem confined to bookshelves and period dramas. The observations, nuances and interplays so vividly drawn in *I Bought a Mountain* may seem almost twee in their portrayal of a now distant past. However, what agriculture of all types (but particularly when one farms in harsher, less fertile environments) teaches those of us who are enrolled in its lifelong apprenticeship, is that the past is never more than a shallow furrow away, its effects as present to our daily work as a video-call with a financial advisor or the switchback changes of national trade and land-use policies.

Everything from the boundaries or limits of a farm's land, through the choice of stock best suited to the grazing and topography, to the underlying fertility of our soils and the vegetation that grows on them is the result of the layered histories of our landscape, histories that span millennia, not decades or economic cycles. Our flocks and their hefts are substantially built on foundations laid down by the Cistercian monasteries who rationalised and controlled the early medieval wool-trade, a pattern which later influenced farm and estate planning post-Reformation. These foundations were themselves built on earlier footings, the sheep-walks and cultivated lands of a Bronze-Age and Romano-British landscape, underpinned yet again by the patterns of cultivation, shepherding, transhumance and migrations of specially-selected and selectively-bred Neolithic livestock, and their farmers. Upland farming always has been, and still is a dance between the needs and limits of the land and environment, and the needs and limits of those who inhabit and work with it, and their animals. Each influences the other, and although the dance changes tempo, evolves and develops, the basic moves are timeless, and the melody it's danced to keeps returning to a common and ageless theme.

The enduring theme of Welsh hill-farming's dance is resilience, it is absolutely central to everything we do. On the western elevations of our landscape, where we farm, not only do our livestock have to contend with far greater seasonal variability in fodder availability than, say, a valley-bottom holding somewhere far inland, but they have to survive, indeed thrive, on coarse grasses, shrub and heather while contending with the perpetual inclemency, highs and lows of the Atlantic weather systems that dominate our climate and working lives in ways unimaginable even to those living in the comparatively close rain-shadow of the Welsh/English Marches. The rain, the extremes in weather – bright sunshine and warm breezes one hour to a near hypothermic soaking the next – described at Firbanks' Dyffryn aren't just familiar punctuation marks for the reader in an unfamiliar narrative landscape, they are explorations of a subject as central to the day-to-day life of an upland farmer and their flock as the price of crude oil is to global financial markets.

The influence of global markets, both financial and commodity, are now being felt in upland Wales in a manner and at a scale entirely unprecedented in agriculture's long history. From the trade in fossil fuels and food to the inflationary forces on land and property prices, external pressures have been building throughout the last century, and the counterpoint they provide to the fluid meter of hill-farming's melody sometimes looks dangerously close to overwhelming the dancers. While public discourse becomes preoccupied with analysing the effects of these pressures, the changes that lie at the root of them often go unexplored or misunderstood. Many of the extensive flocks that typified farms like Firbanks'

Dyffryn are now shadows of their former selves, hefts eroded by a thousand cuts and the deep wound of the Foot and Mouth epidemic; hillsides and plateaus left ungrazed for most if not all of the year, and other areas densely packed in response. The decline of traditional upland cattle breeding mirrors the story of their ovine counterparts. The shepherded migration of livestock uphill in summer, relieving the pressure on lower-lying holdings and pastures, has been substantially arrested by the demand for uniform and fast-grown, efficient meat. Direct employment in hill-farming has fallen significantly, although its indirect effects in the rural economy are still central to the survival of most upland communities. Demographic shifts, tourism, employment and development opportunities, house and land prices have all played their part in reshaping, and potentially eroding the cultural foundations of our agricultural communities – for a community in which Thomas and Esme were two of only a handful of non-Welsh speakers in the 1930s, there is now a palpable sense of fear in Capel Curig that the Welsh language as the medium of community interaction is rapidly being lost. Even the mountains themselves are changing, with greenhouse gas emissions and the effects of man-made climate change contributing to rapid and seemingly irreversible changes in vegetation, soil-structure and hydrology. Where do hill-farming and the communities at its heart fit in this new normal? Is decline the inevitable partner of progress?

Progress has no single or agreed definition, if by progress what we mean is continued exponential growth and the onward march of bland uniformity in all aspects of life, I fear the decline of Welsh upland communities is almost inevitable. Firbanks' description of this process is one of the best I've read, 'The farmers are sacrificed on the shabby altar of industry time after time as the industrialists throw good money after bad in efforts to thaw frozen foreign assets or to stimulate declining barter.' A sentence that has an increased prescience in light of recent changes to the UK's geopolitical positioning. If however what we mean by progress is moving towards a state of sustainable prosperity, living within the limits of our environment and increasing our resilience to what may come in the face of an increasingly volatile future, the deep knowledge embedded in the land and communities of the Welsh uplands has a significant role in informing and driving wider change. Rather than a quaint periphery, viewed from urban and financial centres as little more than the backdrop for memorable holidays, these places, their people and culture become central to our collective efforts to envisage different ways of being.

As with our own farm at Tymawr, if we unpeel the thin veneer of modernity from Dyffryn, the sage green paint and the brightly-anoraked wonderers, lying just beneath the surface are the ready pipelines of sustainable prosperity, power-houses, although dormant, waiting to fulfil their primary purpose. The scattered sheep browsing the rushes are the same bloodline that were graded at Firbanks' Valuation, the culmination of some 6,000 years' breeding and adaptation, as resilient in a changing and changeable climate as the rocks themselves, and capable of producing food and fibres far more sustainable than the packet meals cooked on countless camping stoves, or the thermal base layers of Tryfan's transient climbers. With careful productive management, and with appropriate livestock to justify their cultivation, neglected hay meadows can again become abundant in flowers, humming to the buzz of upland bees. The maintenance of the ffridd's shelter with it's scattered trees, essential in the bright sun of a summer's drought as well as the driving rain and snow of a harsh December, can again form part of a sustainable livestock system, one that evolved long before the coining of words like 'agroforestry'. By prioritising resilience over efficiency, sustainability over growth, the future for Dyffryn and Tymawr alike can and should be so much more than a living museum used to offset the guilt of limitless consumption. For this to become reality, if we're truly genuine about our desire to leave a better world for our children, the real change doesn't need to happen in Capel Curig or Ysbyty Cynfyn, but in the cities, financial, political and industrial centres of the UK and beyond, in the temples of the industrialist's 'shabby altar,' and the hearths, wardrobes, fridges, and thermostats of millions of homes.

THOMAS FIRBANK (1910 – 2000) was a Canadian/Welsh author, farmer, soldier and engineer. He enlisted during the Second World War and was awarded a Military Cross. After his marriage to Esmé Cummins ended, he gave her the Dyffryn farm.

He only returned to Snowdonia in 1993 after a spell living in the Far East, and died in Llanrwst, North Wales.